WORKER SAFETY UNDER SIEGE

WORKER SAFETY UNDER SIEGE

Labor, Capital, and the Politics of Workplace Safety in a Deregulated World

Vernon Mogensen, Editor

M.E.Sharpe
Armonk, New York
London, England

Library of Congress Cataloging-in-Publication Data

Worker safety under siege : labor, capital, and the politics of workplace safety in a deregulated world / [edited by] Vernon Mogensen.
p. cm.
Includes bibliographical references and index.
ISBN 0-7656-1448-0 (hardcover : alk. paper) — ISBN 0-7656-1449-9 (pbk. : alk. paper)
1. Industrial hygiene. 2. Industrial safety. I. Mogensen, Vernon

HD7261.W67 2005
363.11—dc22 2005010491

Printed in the United States of America

The paper used in this publication meets the minimum requirements of
American National Standard for Information Sciences
Permanence of Paper for Printed Library Materials,
ANSI Z 39.48-1984.

∞

BM (c) 10 9 8 7 6 5 4 3 2 1
BM (p) 10 9 8 7 6 5 4 3 2 1

For those who have been injured,
made ill, or killed
while trying to make a living.

Contents

Tables and Figures

Tables

Figures

Acknowledgments

This book is the outgrowth of a special issue on occupational safety and health in a neoliberal world that I edited for *WorkingUSA* (Fall 2003). I owe a debt of gratitude to Immanuel Ness, the editor of *WorkingUSA,* for his sage advice and steady support in both endeavors. An edited book is, of course, a collaborative enterprise, so I want to thank each of the contributors for their efforts and ideas. My colleagues at Kingsborough Community College and other campuses of the City University of New York were also supportive. A special note of thanks goes to Frances Fox Piven for her inspiration and support. I have learned much by working with my fellow watchdogs on the Professional Staff Congress of the City University of New York's occupational safety and health committee. Our work is made easier by the dedicated staff of the New York Committee for Occupational Safety and Health, which provides invaluable resources and training for workers and their unions. I would also like to thank the countless rank and file workers, safety activists, and students who have shared their experiences. Writing a book provides an education, and publishing it provides another.

I would also like to thank Lynn Taylor, my editor at M.E. Sharpe, for believing in the importance of this book, and her colleague Amanda Allensworth for shepherding the manuscript through the publication process. Finally, my greatest debt of thanks goes to Carol Comeau.

Introduction

Vernon Mogensen

Publicity is justly commended as a remedy for social and industrial diseases.
—*Louis D. Brandeis,* Other People's Money

In a truly civilized society, protection of the worker should be regarded as the most essential, irreducible aspect of production cost.
—*Rene Dubos*[1]

Pray for the dead, but fight like hell for the living.
—*Mary Harris "Mother" Jones*

The globalization of the free-market economy is eviscerating the sociopolitical framework that assures workers the rights to free association and safety and health protection in the United States and around the world. Corporate interests, their political allies, and state officials are dismantling the social safety net of labor, environmental, and consumer protections painstakingly pieced together by successive generations of grassroots movements during the Progressive Era, the New Deal, and the Great Society of the 1960s. The same pattern can be seen elsewhere around the developed world. Thomas L. Friedman approvingly explains the logic of globalization:

> The driving idea behind globalization is free-market capitalism—the more you let market forces rule and the more you open your economy to free trade and competition, the more efficient and flourishing your economy will be. Globalization means the spread of free-market capitalism to virtually every country in the world. Therefore, globalization also has its own set of rules—rules that revolve around opening, deregulating and privatizing your economy, in order to make it more competitive and attractive to foreign investment. (2000, 9)

However, the process of globalization, also known as neoliberalism, involves the deregulation of the social safety net protecting workers and society, the public, and the environment and the privatization of government services (Greider 1997). The Bush administration is planning to outsource the Occupational Safety and Health Administration's (OSHA) inspections to private businesses, cutting public expenditures for social services for workers and the poor, and reducing the idea of protecting the public interest to "personal responsibility," a euphemism for the nineteenth-century legal doctrine whereby the worker—not the employer—assumes the risk and costs of injury, illness, and death.

The globalization of the free-market ideology justifies the outsourcing of jobs from highly developed countries to less developed countries based on the classical economist David Ricardo's concept of comparative advantage: jobs that can be performed more profitably in low-wage locales should be outsourced from high-wage locales. The *New York Times* reports that India, China, and the countries of the former Soviet Union have average wage advantages of 85 to 95 percent over the United States. According to Mark Gottfredson, a global outsourcing consultant: "There has never been an economic discontinuity of this magnitude in the history of the world" (Rai 2004, W1). N. Gregory Mankiw, President Bush's chairman of the Council of Economic Advisors, was blunt: "[O]utsourcing is just a new way of doing international trade. . . . More things are tradable than were tradable in the past and that's a good thing" (Cassidy 2004, 26). But free-trade economics does not take the social costs of capitalism into account. According to the cold logic of economists, safety and health problems are considered to be *externalities,* not part of the cost of doing business. The outsourcing of jobs from high-wage to low-wage locales has the effect of transferring relatively safe jobs from developed countries, where safety standards and trade unions are well-established, to more dangerous locales like China, where bonded-, prison-, and child labor is abundant, but both safety and health enforcement and free-trade unions are virtually nonexistent. The deaths of 2 million workers annually, including 22,000 child laborers, are attributable to workplace accidents and occupational diseases around the world. Globally, 270 million workplace accidents and 160 million occupational diseases occur each year (ILO 2004, 1).

Corporate interests and free-trade advocates argue that regulation is too costly and cumbersome. Agenda-setting groups like the Business Roundtable claim that businesses need greater flexibility, voluntary compliance, and deregulation of labor laws if they are to compete in the increasingly competitive global market. But the corporate claim that safety regulations are too expensive to comply with and would bankrupt businesses is not substantiated by experience. A 1995 study of OSHA regulations in effect for at least five years by the Office of Technology Assessment (OTA) found that businesses, and even OSHA, routinely overestimated the cost of compliance. For example, the plastics industry opposed OSHA's 1974 vinyl chloride regulation on the grounds that it would cost it between $65 and $90 billion. OSHA's technical consultant estimated it would cost $1 billion, but the actual compliance cost was only in the $228 to $278 million range (U.S. OTA 1995, 10, 89).

Although free-market advocates claim that the benefits of deregulation and massive tax cuts for corporations and the wealthy will trickle down to workers, class stratification and inequality in the United States, which has the worst disparity between rich and poor of any industrial democracy, are growing. Household income inequality shrank from 1947 to 1968, during the heyday of regulation and the social welfare state, but it grew between 1968 and 2001, during the era of deregulation and the dismantling of social welfare programs (Jones and Weinberg 2000; USCB 2004). Further evidence that economic benefits from tax cuts for the corporations and the rich do not trickle down to the middle and working classes and stimulate demand comes from the Congressional Budget Office (CBO), which found that "President Bush's tax cuts have shifted federal tax payments from the richest Americans to a wide swath of middle-class families" (U.S. CBO 2004). Analyzing Bush's and the Republicans' 2001, 2002, and 2003 tax cuts, the CBO found that the top 1 percent of all taxpayers, with average earnings of $1.2 million, were rewarded with an average tax cut of $78,420 in 2004; that is one-third of all three tax cuts combined. The middle quintile of taxpayers, whose average earnings is $51,000, received only a $1,090 cut. Those in the bottom 20 percent, averaging earnings of $16,620, received only a $250 tax cut (Andrews 2004, A1).

Despite corporate claims that globalization benefits working people in developing countries, it has only widened the gap between the rich and poor countries. The World Bank reports that in 1960 the per capita gross domestic product in the twenty wealthiest countries was eighteen times greater than that in the twenty poorest countries. The present trend toward the globalization of free markets, in place since the 1970s, has only helped to widen the gap between the rich and poor; in 1995, the richest countries were thirty-seven times wealthier than the poorest countries (World Bank 2001).

Workers' rights are human rights, but free markets have not always meant freedom for workers. Article 23 of the Universal Declaration of Human Rights establishes every person's right to work without discrimination, the right to fair compensation for one's work, the right to equal pay for equal work, and the right to form and join free-trade unions. But the World Trade Organization (WTO), which sets the rules of global trade, does not take workers' rights into account. The North American Free Trade Agreement (NAFTA) was passed without labor and environmental protections, but it empowers foreign corporations to challenge member countries' domestic laws that limit the use of their products in order to protect safety and health, the environment, and consumer rights as unfair restraints on trade. Plans are in the works to expand NAFTA to cover virtually the whole western hemisphere without rectifying its shortcomings. Moreover, the WTO is seriously considering adopting the same investment rules that give corporations, but not workers, the global right to sue.

In China, where workers do not have the right to form their own unions, workplace safety laws are weak and rarely enforced, and employers flaunt minimum-wage laws. In 2004, the AFL-CIO took the unprecedented step of petitioning the Bush administration and the U.S. trade representative to bring pressure on China to stop violating workers' rights as unfair trade practices. The AFL-CIO's petition describes working conditions in China's export factories:

> They enter the factory system, and often step into a nightmare of twelve-hour to eighteen-hour workdays with no day of rest, earning meager wages that may be withheld or unpaid altogether. The factories are sweltering, dusty, and damp. Workers are fully exposed to chemical toxins and hazardous machines, and suffer sickness, disfiguration, and death at the highest rates in world history. They live in cramped cement-block dormitories, up to twenty to a room, without privacy. They face militaristic regimentation, surveillance, and physical abuse by supervisors during their long day of work and by private police forces during their short night of recuperation in the dormitories. (AFL-CIO 2004a, 2)

The AFL-CIO's petition charged that if China enforced its own laws, production costs would be 44 percent higher on average (AFL-CIO 2004a). The petition was filed under Section 301, a 1988 amendment to the Trade Act of 1974, which considers workers' rights violations to be unreasonable trade practices. While presidents, including George W. Bush, have routinely supported Section 301 complaints filed by corporations to protect their intellectual property rights, he, consistent with his antilabor policies, rejected the AFL-CIO's petition.

The logic of a global free market encourages a relentless economic race to the bottom, whereby countries continually compete with one another to create a better business environment by offering subsidies to lure new investments, cutting tariffs, deregulating labor laws, ignoring human rights violations, and weakening safety and health protections. This race to the bottom affects countries like China, but it also affects industrial democracies like the United States, where workers' rights and safety protections are under assault. As if its massive tax cuts for the rich and corporations were not enough, President Bush and the Republican-controlled Congress have all but declared class war on America's workers. They have repealed OSHA's ergonomics standard and stripped basic labor rights from 170,000 federal workers who were reorganized into the Homeland Security Department. As if to prove the critics correct in their beliefs that social mobility for the working class is more myth than reality, Bush and Congress have refused to raise the federal minimum wage and plan to amend the Fair Labor Standards Act to make more than 6 million workers ineligible for overtime pay (Eisenbrey 2004, 1).

Acting unilaterally, the Bush administration has repealed the requirement that employers keep records of musculoskeletal injuries, and it has withdrawn numerous proposed safety standards in the pipeline at OSHA and the Mine Safety and Health Administration (MSHA), including the standard to prevent the contraction of tuberculosis in the workplace, the standard on exposure to glycol ethers, and even the rule requiring employers to pay for personal protective equipment that, as several of this volume's essays will explain, is vital to the safety and health of low-wage and foreign-born workers who do many of the most hazardous jobs. The Bush administration is on course to become the first administration in OSHA's history to not issue a single safety and health standard. They have turned the rule-making philosophy of preventing occupational injuries and illnesses on its head and are killing pending rules to prevent them from being promulgated; through 2003, twenty-four of forty-four pending OSHA rules were killed, eighteen of them in 2001 (Goldstein and Cohen 2004, A1). At the same time, the Bush administration expanded employer voluntary programs at the expense of enforcement, and established partnerships with businesses that exclude workers and their unions. Every year since taking office, this administration has tried to cut the OSHA, National Institute for Occupational Safety and Health (NIOSH), and MSHA budgets, slash OSHA training programs by 75 percent, and eliminate funding for union-run programs. In the only bright spot for organized labor, it has been able to prevent these cuts from being fully made. Still, OSHA is so underfunded and understaffed that it would take 106 years for its inspectors to visit all the workplaces under its jurisdiction (AFL-CIO 2004b). This represents a dramatic three-decade decline in OSHA's ability to enforce compliance with safety standards; in 1974, it only would have taken seventy-seven years (Smith 1976, 62). The Bush administration's latest initiative is to strip NIOSH of its relative autonomy by demoting it within the Centers for Disease Control's administrative hierarchy. The Bush administration actively solicits the support of working Americans on social and cultural issues, but is conservative with its compassion when it comes to ameliorating their economic plight. When a reporter asked a question about the surfeit of low-paying, dead-end jobs being created, Susan Sheybani, assistant to the Bush campaign's spokesman Terry Holt, said: "Why don't they get new jobs if they're unhappy—or go on Prozac?" (Reuters 2004, 1).

If fully implemented, globalization of the free-market gospel will return us to the political environment of laissez-faire capitalism that prevailed during the nineteenth century. Injured and diseased workers were considered expendable; little—if any—provision was made for the support of families of workers killed on the job and effective governmental safety regulations were virtually nonexistent. Karl Marx described how the division of labor under industrial capitalism exacerbated workplace safety and health problems:

> Some crippling of body and mind is inseparable even from the division of labour in a society as a whole. However, since manufacture carries this social separation of branches of labour much further, and also, by its peculiar division, attacks the individual at the very root of his life, it is the first system to provide the materials and the impetus for industrial pathology. (1976, 484)

Faced with a rising labor and socialist grassroots movements calling for the eight-hour day and the strict regulation of workplace safety, some corporations began to understand that support for a modicum of regulation was in their interest and allowed them to influence the outcome of the public policy-making process (Weinstein 1968). Consequently, the call among Progressive Era reformers was to put "safety first!" While safety was not put first over profits, the first important regulatory steps toward protecting workers' safety were taken during the early decades of the twentieth century. Many states passed workers' compensation laws and measures mandating that guards be placed on dangerous machinery, and fire-safety codes were adopted in the wake of the infamous Triangle Shirtwaist Fire. The labor movement's pressure during the New Deal resulted in the passage of the National Labor Relations (Wagner) Act, the Fair

Labor Standards Act, and the little-known Walsh-Healy Public Contracts Act. The latter required that employers with federal government contracts worth more than $10,000 meet basic safety requirements.

But further progress was stalled until the 1960s when once again a safety movement arose within organized labor and lobbied for universal national regulations to replace the ineffective, patchwork quilt of state and federal regulations. The passage of the Occupational Safety and Health Act (OSH Act) of 1970, which assured a safe and healthful workplace for all working men and women, was a landmark moment for the labor movement and for efforts to attain a "truly civilized society." It created OSHA to establish and enforce national standards and NIOSH to conduct medical scientific studies of workplace hazards and make recommendations on standards to OSHA. The OSH Act was part of an impressive array of social regulatory legislation passed during the late 1960s and early 1970s in the areas of environment, consumer protection, welfare, and expanded social security provisions that created a safety net to provide Americans protection against the social costs of the market economy.

Nevertheless, many policy makers, supported by corporate contributions and arguments, have uncritically accepted the arguments for deregulation. As a result, OSHA and NIOSH have been chronically underfunded and understaffed since their inception in 1971. With social amnesia regarding all that society has historically learned about the necessity of regulating capitalism's excesses, public policy makers are marching forward into the nineteenth century of market economics.

Given this global push toward deregulation and the rollback of worker rights, how committed are corporations and governments to putting safety first? This volume is divided into three parts: part I: Free-Market Ideology and the Evisceration of Workers' Safety Rights; part II: Old and New Challenges to Occupational Safety and Health in the United States; and part III: The Impact of Neoliberalism on Workers' Safety Rights Abroad. The first two chapters in part I focus on the social and legal aspects of the double standard that apply to death in the workplace. Causing the reckless death of a person outside the workplace is a felony, but willfully ignoring hazardous conditions that cause a worker's death on the job is only a misdemeanor. It is rare for OSHA, or state prosecutors, to seek legal punishment for companies whose willful acts result in worker deaths; and when they do, the punishment is less than for violations of environmental or financial rules. A recent analysis by the *New York Times* found that "since 1990, [OSHA] has quietly downgraded 202 fatality cases from 'willful' to 'unclassified,' a vague term favored by defense lawyers in part because it virtually forecloses the possibility of prosecution" (Barstow 2003, A1).

In chapter 1 Jordan Barab analyzes the problem of why our society treats the workplace deaths of some more seriously than others. In a sense, the tremendous improvements in workplace safety over the past ninety years seem to have made deaths on the job less remarkable. However, Barab points out how we take our cues from the mass media, which highlight the deaths of astronauts but virtually ignore the deaths of day laborers or construction workers. Barab also observes that the most expendable Americans—immigrants and the poor—do some of the nation's dirtiest and most dangerous work, and might be considered less expendable if both media and society treated the dangers they face with greater respect. He calls on the labor movement and the safety and health community to confront this problem head-on and develop an educational strategy to rectify the distorted idea that some lives are more valuable than others.

In chapter 2 Rory O'Neill investigates how employers around the world go unpunished even when clear negligence and cost cutting on their part contribute to the deaths of their workers. He argues that criminal prosecutions are rarely brought in these cases, even when prosecutors have the probable cause evidence necessary to prove their case beyond a reasonable doubt. As a result, negligent employers can almost always kill with impunity. O'Neill reports on the global campaign by unions in the United Kingdom, the United States, Australia, and elsewhere around the world who are now calling for corporate accountability to extend to workplace safety crimes, and for the incarceration of the most dangerous offenders. In June 2003 U.S. senator Jon Corzine (D-NJ) introduced federal legislation that would change the charge from a misdemeanor to a felony and increase the maximum criminal penalty from six months to ten years in prison for those who willfully violate workplace safety laws and cause the wrongful death of an employee.

In chapter 3 Peter Dorman asks whether expert paternalism is the answer to worker irrationality. Mainstream economics has long assumed that people are rational actors who make decisions based on their own interests. It presumes that employees in hazardous workplaces freely assume the risks of injury, illness, and death, or else they would quit their jobs. But Dorman notes that economists have rediscovered psychology under the rubric of "behavioral economics." Workers who lack portable skills or higher education, and who have bills to pay, may not have the option of quitting a dangerous job. Moreover, workers may not be aware, or be informed by the employer, of the dangers that confront them. While there are aspects of rational methods that are meritorious, Dorman argues that the "rational/irrational framework is profoundly misleading." In addition to setting permissible exposure limits, risk regulation should be concerned with greater information sharing, joint occupational safety and health committees, right-to-know laws, and trust building between workers and management. Finally, he maintains that giving workers greater voice in decision making and forming stronger unions are necessary to ensure balance in collective bargaining and reduce the pressures on workers to take dangerous risks.

Part II examines old and new challenges to the maintenance of occupational safety and health in the United States. In chapter 4 Gerald Markowitz and David Rosner analyze the continuing battle against the scourge of silicosis that they first analyzed in their 1991 book *Deadly Dust.* In that book they tracked the decades-long rise of this deadly occupational disease that affected thousands of lives and its seeming disappearance from public awareness in the 1950s and 1960s. After publication of *Deadly Dust,* they began to receive calls from trial lawyers asking them to appear as expert witnesses at trials for a new generation of workers with silicosis. During the pretrial "discovery" process lawyers for the workers uncovered thousands of documents that detailed the collusion between corporate interests and government officials to suppress efforts to tighten occupational safety standards for silica exposure. This case study demonstrates the institutional weakness and lack of autonomy from elite pressure of OSHA, NIOSH, MSHA, and the Environmental Protection Agency, which in theory should have been looking out for workers' safety and the public interest. During the Carter administration, the silica industry's ironically named Silica Safety Association succeeded in blocking OSHA's efforts to promulgate a silica safety standard that would have banned the use of sand as an abrasive. In 1997 history repeated itself when industry formed the Silica Coalition to block expected efforts by OSHA to once again promulgate a standard. But OSHA still has not done so. The only recourse for diseased workers has been to file civil lawsuits, but corporate interests are also trying to eliminate this civil right under the guise of "tort reform."

As the U.S. economy continues to integrate into the global economy, we have become increasingly reliant on foreign-born workers at the expense of their safety and health on the job. Foreign-born workers increased 22 percent from 1996 to 2000 but fatal injuries among foreign-born workers increased by 43 percent during the same time period (Loh and Richardson 2004, 42). Foreign-born workers do many of society's unwanted and dangerous jobs and they are faced with cultural and language barriers. In chapter 5 Laura H. Rhodes examines the subculture of marginalized Hispanic workers who speak little or no English and must cope with safety manuals and warning signs they cannot read and oral instructions they do not understand. Hispanics account for 15 percent of all worker fatalities, 25 percent higher than the rate for all other workers. With the Hispanic workforce projected to surpass the African American workforce in size by 2008, this problem is expected to continue to grow.

But Rhodes argues that the magnitude of the problem is even larger, because the U.S. Bureau of Labor Statistics' data do not include the illegal or undocumented workers. Moreover, Hispanics are less likely to have health care coverage and the health care they do have access to is more likely to be of inferior quality. They are also more likely to live in polluted communities, raising environmental justice issues, and these disparities are most acute along the U.S.–Mexico border. It seems that little has changed for immigrant workers in a century's time; 146 immigrant workers were killed in the Triangle Shirtwaist Fire of 1911 largely due to locked doors. In 1991, twenty-five workers died under similar circumstances at Imperial Foods in Hamlet, North Carolina. Reports of Wal-Mart and grocery store workers being locked in without a key on overnight shifts raises concern that this tragedy could repeat itself.

Historically, safety and health protection has focused on imminent hazards in the industrial workplace. But as the next two chapters in part II demonstrate, white-collar and service-sector workers, who now represent the overwhelming majority of American workers, increasingly face serious safety and health hazards that need to be addressed. Joan Greenbaum and David Kotelchuck write about the ubiquitous problem of poor indoor environmental air quality in chapter 6. OSHA started working on an indoor air quality rule in 1994. As late as April 2001, OSHA said that "every day . . . 20 million American workers face an unnecessary health threat because of indoor air pollution in the workplace. Thousands of heart disease deaths, hundreds of lung cancer deaths, and many cases of respiratory disease, Legionnaire's disease, asthma, and other ailments are estimated to be linked to this occupational hazard" (OMB Watch 2004, 3). Despite this widespread threat to public health, President George W. Bush's team came to the astonishing conclusion that there was no need for an OSHA rule because many state and local governments and private employers had already tackled the environmental smoke problem. But this was not a decision based on sound policy. A patchwork quilt of state and local government rules was no substitute for a federal rule, and the Bush administration simply dismissed other sources of indoor air pollution. It announced that "The agency found that withdrawal of the proposal would allow it to devote its resources to other projects" (OMB Watch 2004, 3). Despite the well-documented health problems associated with indoor air pollution, the Bush administration sided with the restaurant industry, which contributed over $1.2 million to kill the rule (Pickler 2004, 2). The majority of requests for NIOSH health hazard evaluations involve poor office environmental quality, up from only 8 percent in 1980. This worsening trend did not stop the Bush administration from killing OSHA's pending rule to prevent restaurant and other workers from being exposed to tobacco smoke and other indoor air pollutants in December 2001. Their Catch 22 rationale was that state and local rules had made OSHA's rule unnecessary. Of course that is not true and consequently unions are becoming increasingly active on the issue. Greenbaum and Kotelchuck, who co-chair the safety and health committee of the Professional Staff Congress at the City University of New York (CUNY), describe the innovative grassroots approach the union is taking with its "Got Air?" campaign throughout the CUNY system. They note that "experience and activism combined, particularly within the crucible of union activity, is a potentially powerful way to both remedy immediate problems and to set the agenda for further research."

Chapter 7 explores the continuing struggle over the growing problem of work-related musculoskeletal disorders (WMSDs), which include repetitive strain illnesses. The problem has dramatically increased with the spread of repetitive motion and computer-automated work in both blue-collar and white-collar sectors of the economy. WMSDs now represent the leading occupational safety and health hazard, but many of them can be prevented through the science of ergonomics. The struggle for an ergonomics standard presents a good opportunity to test the theories that government administrators develop programs to build state capacity and autonomy from societal pressures. Instrumentalist and structuralist theories shed more light on the fact that socioeconomic interests were key in determining the ergonomics standard's fate. The impetus for the ergonomics standard came from a coalition of organized labor, women's groups, and committees on occupational safety and health, not OSHA. Labor fought for two decades to secure the ergonomics standard, but sustained pressure from corporate interests through their policy-planning network, especially after Republicans won control of Congress in 1994, forcing the Clinton administration to make numerous concessions. The corporate campaign to discredit the science of ergonomics and the WMSD crisis delayed the standard's release until late 2000. The Republican-controlled Congress and President Bush used an obscure law, the Congressional Review Act, to repeal it, the first time an OSHA standard has ever been repealed. While organized labor demonstrated that it still has the capacity to build and sustain coalitions to achieve and protect gains through political action, this case study shows the dominant role of corporate interests in the making of public policy. The attack on, and defeat of, the ergonomics standard is part of a larger corporate effort to "reform OSHA," a euphemism for replacing regulation with voluntary compliance. The latter approach gives corporate interests greater leverage to discipline the labor force, limit workers' rights, and further unravel the social safety net in the name of maintaining global competitiveness.

Part III contains case studies on the globalization's impact on safety and health outside the United States. The first two reports are from Canada. They examine the devastating consequences of the probusiness Progressive Conservatives coming to power during the 1990s and their dismantling of worker rights and safety protections that had been constructed over the years by the New Democratic Party and the Liberals. In chapter 8 Penney Kome examines the interconnections between gender, nationality, and occupational health. Her contribution, "The 10 Percenters," carries a double meaning. First, she reports that 10 percent of adult Canadians, mostly women, have debilitating WMSDs that seriously limit the scope of their daily activities. Second, for many years they were forced to estimate how many of their fellow workers had occupational illnesses like WMSDs by dividing the U.S. Bureau of Labor Statistics (BLS) figures by a factor of ten. Canadians have a distinct disadvantage in fighting occupational injury and illness since record keeping was outsourced by the government during the 1990s to thirteen regional Associations of Workers Compensation Boards of Canada, a nonprofit organization that charges fees to access the data that were formerly available to the public for free. Moreover, it only tabulates cases for which lost-time claims are approved, thereby understating the true magnitude of the occupational injury problem in Canada. Consequently, Canadians are kept in the dark about the scope of the repetitive strain injury (RSI) problem, thereby limiting the ability of labor and society to identify the extent of the problem and combat it.

In chapter 9 Robert Storey and Eric Tucker provide a historical analysis of worker participation and occupational health and safety regulation in Ontario from 1970 to 2000 in light of the rise of neoliberal policies. They describe a shift from systems of mandated partial self-regulation in which workers had the right to participate, supported by external enforcement of regulations, to more ambiguous models that included the downsizing of government and voluntary compliance by employers. In the 1990s the Progressive Conservatives instituted a plan to roll back the safety net of occupational safety regulations and other worker protections that were the achievement of the defeated New Democratic Party. They observe that Ontario's deregulation of occupational safety and health and the weakening of workers' rights to participate in the policy-making process was not an isolated occurrence. It was part of the global trend to conform to deregulatory polices that are part and parcel of neoliberalism.

In chapter 10 Carlos Eduardo Siqueira and Nadia Haiama-Neurohr look at free-market policies' deleterious effects on safety and the environment in Brazil through the case study of the P-36 oil-drilling platform. In March 2001 a series of explosions sank the Petrobrás oil company's P-36 platform in the Atlantic Ocean with the loss of eleven lives. The sinking prompted a national debate over Brazil's adoption of neoliberal policies at the expense of workers' rights and livelihoods. Siqueira and Haiama-Neurohr ask why the P-36 platform sank. Who was to blame for such a catastrophic disaster? Could the platform have been salvaged after the first explosion? Why did eleven workers die? They argue that the P-36 platform's sinking was a consequence of Petrobrás's organizational strategy to cut safety and labor costs, replace skilled workers with lower paid unskilled workers, lax maintenance, and increase production in order to maximize profits in the global free-market economy. This pressure was intensified by the need to raise oil revenues for the Brazilian treasury. These neoliberal policies were adopted by Brazil and much of the developing world a decade after they were adopted in the United States and Europe. Longtime trade unionist and political activist Luis Inácio Lula da Silva was elected to the presidency in 2002 on the platform of rolling back the worst excesses of neoliberalism, and he did institute labor and environmental reforms at Petrobrás. But free-market policies are so deeply embedded at both Petrobrás and in the world economy that the scope of his near-term options may be limited. Still, Siqueira and Haiama-Neurohr note that it is too early to evaluate the success of Lula's efforts to reinstitute safety and environmental reforms.

With the fall of communism during the period 1989–91, market economists and policy makers were eager to transform Eastern Europe into prosperous market economies with high employment and equality of opportunity for all. However, Russia has experienced increasing inequality, high unemployment, and growing unrest over poor working conditions. The wealthiest 20 percent of Russians account for 53.7 percent of the country's income, while the poorest 20 percent earn only 4.4 percent (World Bank 2001, 51). As Michael Haynes and Rumy Husan point out in chapter 11, perhaps nowhere in the world have free-market policies been instituted more quickly than in the countries of the former Soviet bloc. Using Russia and Hungary as their comparative case study, they inquire as to how workers' health and safety fared under policies of market reforms and deregulation. While Hungary has had greater success than Russia in adapting to the new economic realities of globalization, the illusion of workers' safety has survived from the days of the old regimes, albeit in a new form. In both countries workers suffer from lack of enforcement of safety laws, underreporting of accidents, powerful employers, and weak unions. Haynes and Husan conclude that while the impact of market reform has differed in Russia and Hungary, the introduction of neoliberalist policies have not been a panacea for the occupational safety and health problems that prevailed under communism.

Organized labor is on the defensive in the United States and elsewhere around the industrialized world. Union membership has declined from its high-water mark of 35 percent in 1954 to only 12.9 percent of the workforce in 2003; private sector union membership is even lower, at 8.2 percent, levels not seen since the 1920s (Century Foundation 1999, v; BLS 2004, 1). Employers who coerce, intimidate, and fire workers for trying to form unions or for complaining about safety violation, are rarely punished despite the fact that these tactics are forbidden by both U.S. law and the Universal Declaration of Human Rights. Labor law does allow employers to hold "captive audience" meetings where workers are forced to listen to antiunion messages. If workers' organizing efforts overcome these obstacles, employers can delay and appeal outcomes of certification elections for years, refuse to negotiate with the union the workers chose to represent them, or close the workplace and move elsewhere. Sanctions are so insignificant that they do not pose a deterrent to offending employers. A bill in Congress, the Employee Free Choice Act, would impose stiffer penalties against employers who violate workers' right to organize, and would dispense with the need for a certification election if a majority of workers signed cards favoring unionization. Approximately 32 million workers are unable to organize and collectively bargain under labor law and National Labor Relations Board decisions because of occupational exclusions, including farm workers, domestic workers, independent contractors, and low-level managers. The lack of respect for workers' rights at home weakens the U.S. government's credibility when it tells other countries to respect "core labor standards" (Compa 2000, 2).

Urgent action is needed on the part of the labor movement if it is to be revived. At a meeting to form an executive committee to address the problem of declining membership, John Wilhelm, president of the Hotel Employees and Restaurant Employees, said: "My view is that if we don't devote the largest possible amount of money to organizing and to political action that relates to organizing, we will go out of business." Times are so tough for labor that Wilhelm even considered eliminating the AFL-CIO's Safety and Health Department. "And if we go out of business, we can't help anybody's health and safety" (Greenhouse 2003, 22). But it would be a false economy since workplace safety and health issues have historically been key issues involved in organizing drives. Globalization of the free market is not an inevitable, irreversible force of nature, and it is not universally applied. There are numerous exceptions to the rule; countries still covet their tariffs—witness Bush's willingness to set aside his own free-trade "principles" and international trade law to impose tariffs on imported steel to gain a political advantage in steel-producing states. A coalition of workers and their unions, women's groups, and others, needs to organize and put grassroots pressure on the Democratic Party as they did during the Progressive, the New Deal, and the Great Society eras. Committees for occupational safety and health can play an important supporting role as clearinghouses of information.

U.S. labor needs to increase its efforts to coordinate with labor movements in other lands. In a symbolic expression of global solidarity, April 28 has become internationally recognized as Workers' Memorial Day to remember those who have died on the job. The World Summit on Sustainable Development serves as a counterforum to the WTO meetings, where the promotion of safe and sustainable workplaces is discussed. Sustainable development is the idea that the present generation should balance the harm done by current economic development projects with concern for protecting the environmental and social resources for future generations, including safe and healthful working conditions. The International Confederation of Free Trade Unions, which represents 148 million workers through its 233 affiliated national trade unions in 152 countries and territories, is actively involved in integrating sustainable development concepts with protection of workers' safety and health and workers' overall well-being. The precautionary principle is another conceptual advance from the environmental movement that is gaining acceptance in Europe and North America. It states that producers have an ethical responsibility to demonstrate that their products, such as toxic chemicals, and hazardous substances, such as asbestos, do not pose a threat to human health and the environment before introducing them into the workplace and the broader community.

Workplace safety cannot be separated from the need to protect the environment. The 1984 leak of toxic chemicals from the Union Carbide Corporation's plant in Bhopal, India, stands as the world's worst industrial disaster and a reminder of how the welfare of workers and the community are inextricably linked. While initial estimates were that 3,000 were killed, twenty years later the number of deaths had grown to more than 15,000 from related diseases (Bidwai 2004), and an estimated half-million others were affected by the toxic cloud of gas that blew over Bhopal (Lapierre and Moro 2002, 376–77). Bhopal was labeled an "accident" in the mass media, but it was the result of management cutbacks in safety and quality control in order to maximize profits. The safety systems, which kept the plant's methyl isocyanate stable and contained, were turned off to save a relatively small sum of money when the plant was not in production (Lapierre and Moro 2002, 210–14).

Historically, efforts to improve working conditions have started with the workers themselves, not the ruling elites. The surviving victims of Bhopal and their families banded together to seek justice in the Indian courts. After twenty years of struggle, they overcame numerous obstacles and opposition from powerful groups including the courts, the Indian government, and Union Carbide to win a symbolic court victory. Though their monetary compensation was small, through organization, persistence, and struggle the Bhopal survivors succeeded in keeping the issue focused on corporations and governments that perpetuate a lower standard of safety and health for those in the developing world. Workers must now build on these efforts to link with their peers around the world. A gospel of free markets that does not take social costs into account should not be accepted on blind faith; it must be tempered by the regulatory provision for workers' rights to free association and safety and health.

Note

1. Quoted in statement of Joseph A. Dear, assistant secretary of Labor for Occupational Safety and Health, before the Subcommittee on Labor Standards, Occupational Safety and Health Committee on Education and Labor, U.S. House of Representatives, February 10, 1994.

References

AFL-CIO. 2004a. *Section 301 Petition of the American Federation of Labor and Congress of Industrial Organizations*. Before the Office of the United States Trade Representative. March.

———. 2004b. *Death on the Job: The Toll of Neglect: A National and State-By-State Profile of Worker Safety and Health in the United States*. 13th ed. Washington, DC: AFL-CIO, April.

Andrews, Edmund L. 2004. "Report Finds Tax Cuts Heavily Favor the Wealthy." *New York Times*, August 13.

Barstow, David. 2003. "U.S. Rarely Seeks Charges for Deaths in Workplace." *New York Times*, December 22.

Bidwai, Praful. 2004. "Slow-Motion Justice for Bhopal." *Navhind Times*, July 28.

Cassidy, John. 2004. "Winners and Losers: The Truth about Free Trade." *New Yorker*, August 2: 26–30.

Century Foundation. 1999. *What's Next for Organized Labor? The Report of the Century Foundation Task Force on the Future of Unions*. New York: The Century Foundation Press.

Compa, Lance. 2000. "Unfair Advantage: Workers' Freedom of Association in the United States under International Human Rights Standards." New York: *Human Rights Watch Report*, August.

Eisenbrey, Ross. 2004. *Longer Hours, Less Pay: Labor Department's New Rules Could Strip Overtime Protection from Millions of Workers*. Washington, DC: Economic Policy Institute, July.

Friedman, Thomas L. 2000. *The Lexus and the Olive Tree*. New York: Anchor Books.

Goldstein, Amy, and Sarah Cohen. 2004. "Bush Forces a Shift in Regulatory Thrust: OSHA Made More Business-Friendly." *Washington Post*, August 15.

Greenhouse, Steven. 2003. "Worried About Labor's Waning Strength, Union Presidents Form Advisory Committee." *New York Times*, March 9.

Greider, William. 1997. *One World, Ready or Not: The Manic Logic of Capitalism*. New York: Simon and Schuster.

International Labor Organization (ILO). 2004. *Facts on Safe Work*. Geneva: International Labor Organization.

Jones, Arthur F., Jr., and Daniel H. Weinberg. 2000. *The Changing Shape of the Nation's Income Distribution, 1947–1998*. Washington, DC: U.S. Census Bureau.

Lapierre, Dominique, and Javier Moro. 2002. *Five Past Midnight in Bhopal: The Epic Story of the World's Worst Industrial Disaster*. New York: Warner Books.

Loh, Katherine, and Scott Richardson. 2004. "Foreign-born Workers: Trends in Fatal Occupational Injuries, 1996–2001." *Monthly Labor Review* (June): 42–54.

Marx, Karl. 1976. *Capital: A Critique of Political Economy*, volume 1. London: Penguin Books.

OMB Watch. 2004. *OSHA Items Withdrawn from Agenda*. Washington, DC: OMB Watch, August.

Pickler, Nedra. 2004. "Democrats Criticize Bush's Corporate Ties." *USA Today*, August 16.

Rai, Saritha. 2004. "Financial Firms Hasten Their Move to Outsourcing." *New York Times*, August 18.

Reuters. 2004. "Unhappy Workers Should Take Prozac—Bush Campaigner." July 29. Available at: www.sourcewatch.org/index.php?title. Susan Sheybani.

Smith, Robert Stewart. 1976. *The Occupational Safety and Health Act: Its Goals and Achievements*. Washington, DC: American Enterprise Institute.

U.S. Bureau of Labor Statistics (BLS). 2004. *Union Members in 2003*. News release, January 21. Washington, DC: U.S. Bureau of Labor Statistics.

U.S. Census Bureau (USCB). 2004. Table IE-6. *Measures of Household Inequality: 1967 to 2001*. Available at: www.census.gov/hhes/income/histinc/ie6.html.

U.S. Congress, Office of Technology Assessment (OTA). 1995. *Gauging Control Technology and Regulatory Impacts in Occupational Safety and Health—An Appraisal of OSHA's Analytic Approach*, OTA-ENV-635. Washington, DC: Government Printing Office, September.

U.S. Congressional Budget Office (U.S. CBO). 2004. *Effective Federal Tax Rates Under Current Law, 2001 to 2014*. Washington, DC: Congressional Budget Office.

Weinstein, James. 1968. *The Corporate Ideal in the Liberal State: 1900–1918*. Boston: Beacon Press.

World Bank. 2001. *World Development Report 2000/2001*. New York: Oxford University Press.

I

Free-Market Ideology and the Evisceration of Workers' Safety Rights

1

Acts of God, Acts of Man

The Invisibility of Workplace Death

Jordan Barab

Mustafa Boyraz, a thirty-four-year-old Annandale, Virginia, construction worker, was crushed to death on January 27, 2003, when a granite slab fell on him at work. Marty Nesbitt of St. Louis, Missouri, was killed on March 15, 2003, when he fell thirty feet from the grandstand roof at Fairmount Park Raceway while measuring part of the roof.

These two weeks were typical in that Boyraz and Nesbitt were joined by more than 100 other workers in the United States who are killed on the job each week; there were almost 7,000 workplace deaths in 2001 (U.S. Bureau of Labor Statistics 2001). In addition to the personal tragedy, workplace deaths and injuries cost this nation hundreds of billions of dollars every year (Liberty Mutual 2003). Although this chapter will focus on fatal occupational injuries, let us not overlook the less recognized deaths of 165 Americans every day from occupational diseases caused by exposure to toxic chemicals, infectious agents, and other hazardous substances.

Like most American workers who never make it home from work at the end of the day, they were "lucky" to get a few inches in their hometown newspapers. These victims of workplace hazards often do unglamorous, dirty jobs on construction sites, roads, and factories. They die alone, only noticed and remembered by their immediate family, friends, and coworkers. We do not hear about them mainly because they are just regular people and most of them die one at a time. Some never even seem to have names, because the names are withheld until the next-of-kin is notified. By the time that happens, the media have often already lost interest.

But the weeks that Boyraz and Nesbitt were killed were not typical. In the week that Boyraz was killed, seven other workplace deaths received days of international headlines. Their names, their faces, their families, and their life stories became familiar to millions, if not billions of people around the world. In that second week dozens of other on-the-job fatalities achieved similar notoriety. Those seven the first week were the astronauts of the space shuttle *Columbia,* and those in the second week were American men and women who died fighting in Iraq.

Of course, the invisibility of most workplace deaths is not just a problem in the United States as a recent article about the workplace carnage in Brazil shows:

> Unfortunately, even as Labour Day approaches, these dramatic figures involving on-the-job deaths, injuries and illnesses simply do not captivate the nation—something statistics on urban violence or traffic accidents accomplish a lot more easily. Crimes, especially involving the rich and famous, and multiple car crashes with numerous deaths always make headlines, even if they are quickly forgotten. Perhaps the problem with work-related accidents is that the victims are generally part of the masses, "unknowns" from the poor side of town. Outside of their families, there is little concern about their welfare in Brazilian society, so their deaths hardly attract any attention at all. (Ferreira 2003)

What determines how much press a workplace death gets? Clearly one factor is the number killed at the same time. The equivalent of a Boeing 747 full of workers are killed on the job in the United States every few weeks. Yet, because most are killed one or two at a time, no one notices.

In one sense, we are victims of our own success. In 1913 the Bureau of Labor Statistics documented approximately 23,000 industrial deaths among a workforce of 38 million, equivalent to a rate of 61 deaths per 100,000 workers. In that environment, it was likely that most people knew someone who had been killed on the job. Today, the workplace death toll has fallen to just under 7,000 deaths each year, a rate of 4.3 per 100,000 ("Achievements in Public Health" 1999). "We don't talk about these people much. Their lives are invisible, far from the media pundits. They're often the immigrants and the poor, those most disposable in our culture," according to Paul Rogat Loeb, chair of the Peace and Justice Alliance, writing in the *Seattle Times* about the disparity between coverage of the space shuttle *Columbia* disaster and more common workplace fatalities (Loeb 2003).

Another factor in determining the public visibility of workplace fatalities is the type of jobs that workers do. Astronauts killed while exploring the frontiers of space and the human imagination will always receive more media attention than a Hispanic construction worker who falls off a roof.

The *Washington Post*'s editors, wondering why the astronauts received so much more attention than the deaths in Afghanistan of four U.S. soldiers in a helicopter crash around the same time as the shuttle disaster, suggested that the astronauts embodied "national aspirations of greatness, and human aspirations to reach beyond ourselves." Their heroism and courage was often noted. They, like the soldiers, gave their lives for their country.

All of this is true. It takes courage to fly into space, knowing the dangers, knowing that you may die far from home. But what about the courage it takes for an immigrant to go to work on a hazardous construction site to feed his family, unable to change his working conditions, and knowing that he may die far from home?

What is the difference between the courage needed to go into space, assured that billions of dollars are being spent to bring every astronaut home alive, versus going down into a deep, unprotected trench, suspecting that one's employer is cutting corners on safety to save a few bucks? Millions of workers go to work every day in this country understanding that this society accepts a certain level of death in the workplace, while it demands 100 percent safety in the space program.

President George W. Bush, in his eulogy for the seven astronauts, was most moving when he spoke to their children:

> And to the children who miss your Mom or Dad so much today, you need to know, they love you, and that love will always be with you. They were proud of you. And you can be proud of them for the rest of your life. (2003)

Yet hundreds of children are left without parents every week due to workplace accidents. Do the children of factory workers or prison guards miss their parents any less than the children of astronauts? And why are "aspirations of greatness" more valued in our society than aspirations of being a good parent and friend and coming home from work safely at the end of the day?

The small number and short length of articles about workplace fatalities make it difficult for Americans to understand the toll that these fatalities take on their families, their coworkers, their communities, and our society. But the number of column inches may not be as significant to America's comprehension of workplace dangers as the content of the articles. How are the causes of the fatalities characterized and what are the implications for worker safety in this country?

Economist John Mayo, dean of the Georgetown University McDonough School of Business in Washington, D.C., and executive director of the Center for Business and Public Policy, approached one explanation in a *Houston Chronicle* column that appeared shortly after the shuttle disaster:

> Why aren't more people concerned? Unfortunately, much of the public thinks accidents are inevitable. The truth is that workplace deaths are especially tragic because most are preventable. Importantly, prevention of injuries and death are often not the product of expensive, massive investments, but rather simply sound management systems and practices. (2003)

If workplace deaths are inevitable, if there is nothing anyone can do about them, then the employer has limited responsibility for workplace safety and the government, as represented by the Occupational Safety and Health Administration (OSHA), has no useful role in issuing or enforcing workplace standards. In an article titled "Abolishing OSHA" the Cato Institute, a libertarian think tank based in Washington, D.C., argued that:

> [The Bureau of Labor Statistics] also found that 40 percent of recent workplace fatalities were from transportation accidents (almost half the fatal transportation accidents were highway accidents), and about 20 percent of workplace fatalities were from assaults and other violent acts (over 80 percent were homicides and 15 percent were suicides). In other words, only 40 percent of workplace fatalities were caused by dangers thought by most to be unique to the workplace, such as the classic example of falling into a machine. *The leading causes of work-related deaths in recent years, transportation accidents and assaults, are unlikely to be reduced much by OSHA inspections.* (Kniesner and Leeth 2000; emphasis added)

But highway fatalities dismissed in this article as unpreventable were not so readily dismissed by the U.S. military, which was alarmed at the number of traffic accident fatalities during the first Gulf War. Since that time, the military has made major—and reportedly successful—efforts to reduce injuries and fatalities resulting from driving accidents. According to an account in the *Washington Post*:

> Compared with that from the first Gulf War, data from the latest fighting also reveal a dramatic reversal in the ratio of combat to noncombat casualties.
>
> Twelve years ago, 50 percent more soldiers died in accidents (235) than in battle (147). In the recent war, there were only a third as many noncombat fatalities (36) as deaths in battle (101). The same pattern

> appears to hold for nonfatal injuries, with the data on evacuated Army troops showing that 107 had noncombat injuries, compared with 118 who had combat wounds.
>
> The Army attributed the steep drop in noncombat deaths and injuries, in part, to the Army's effort to improve driver safety and to ensure that soldiers were well rested when operating vehicles. In the first Gulf War, motor vehicle accidents alone accounted for about half of all serious injuries. "Because this was such a motorized effort, we expected many more accidents than we actually saw. I think this is a definitive success story," [an Army spokesperson] said. (Brown 2003)

And while there is no OSHA regulation addressing the problem of workplace violence, federal OSHA and several state OSHA plans have issued guidelines containing numerous recommendations that have been proven to substantially reduce the number of violence-related workplace fatalities (OSHA 1998). Panic alarms and adequate staffing in mental health institutions, and video cameras and lock-drop safes in all-night convenience stores are just a few measures that have been shown to be effective in preventing "inevitable" fatalities resulting from workplace violence (Barab 1997).

Unfortunately, it is still not uncommon for the public and the media to assume that workers are to blame for many workplace accidents even though it has been thirty years since the Occupational Safety and Health Act gave the employer responsibility for providing safe working conditions. On one level this "blame the worker" philosophy is the result of unawareness. Even union representatives often find it easy to believe in blame-the-worker theories, not because of a disdain for their members, but because identifying the root causes of accidents seems too technical and they have not been trained to identify those causes or preventive measures.

For example, highway construction workers are frequently struck and killed by moving equipment in the work zone. Other than inattentiveness, what can explain a worker walking right into the path of a moving vehicle? Not until one spends time closely observing those who do the job—sometimes for ten hours at a stretch—can one understand how noise, heat, dust, fatigue, work pace, and disorganization lead to a moment of "inattentiveness."

Blame-the-worker theories are nothing new, but they have recently become "legitimized" into the concept of behavioral safety: instead of management taking responsibility for eliminating hazards and providing safe working conditions, blame is placed on workers who make mistakes. Workers are therefore encouraged to "be careful," either by offering incentives for not being injured or by punishing workers who get injured on the job.

In a recent memo to its employees, Sysco Food Services of Baltimore, Maryland, declared that: "All accidents or incidents resulting in accidents that involve work-related injury, damage to equipment, merchandise or other property will be assigned points." Points were also awarded for Workers Compensation claims and lost time. Twelve points leads to "verbal counseling," twenty-four points earns a three-day suspension without pay, and thirty points leads to termination (Sysco 2000).

In the story about the death of Marty Nesbitt that opened this chapter, no mention was made of the fact that legally required fall protection measures had not been provided. In fact, the final paragraph left readers wondering if Nesbitt had not gotten what he deserved:

> A toxicology test showed no alcohol in Nesbitt's system. Madison County Deputy Coroner Ralph Baahlmann said that at some point prior to the accident Nesbitt had used marijuana and that he may have been under the influence of marijuana at the time of the accident. (Horrell 2003)

And an article about a criminal indictment handed down against two California employers for the asphyxiation deaths of two of their employees in a manure pit contained the following curious statement by the owners' attorney: "Neither of these guys [the owners] did anything. They're charged with not doing something," he said. "The poor, unfortunate victims made choices on their own" (Conway 2003).

In some ways, even coworkers feel more comfortable believing the worker was at fault, rather than the working conditions. It is easier to live in denial with the belief that "It could never happen to me because I would never be so careless as that guy," than to understand that unsafe working conditions caused the death of one's coworker—conditions that could just as easily have left one's own children without a parent. If the fault lies with unsafe working conditions and not with individual failure, then one either need to be scared or one has to do something to change the working conditions. Confronting the boss about safety conditions or even requesting an OSHA inspection are not generally comfortable options for American workers, especially for the majority who do not have the protection of a union. For immigrant workers, especially illegal immigrants, such options are generally out of the question.

In addition to blaming workers for workplace injuries and fatalities, employers often blame "acts of God" or the "whims of Mother Nature" to explain such "unpredictable" tragedies as the collapse of a twelve-foot-deep trench on top of a construction worker or the asphyxiation of a sewer worker in an unmonitored confined space. "Who could have predicted it?"

A recent article describing the background to the 2002 Pennsylvania coal mine flood that ended in the rescue of the trapped miners illustrates this point. The owner/operator of the mine, David Rebuck, had received multiple warnings about the inadequacy of their map that showed adjacent mines that had been flooded. Despite these warnings, Rebuck called the flooding an "act of God" in a local television interview. As University of Pennsylvania Professor Charles McCollester wrote:

> The flood of testimonials to the mercy of God threatens to obscure the very human factors that led to the near-disaster. God may well have had a hand in the rescue, but human avarice and more than a century of fierce corporate manipulation and struggle for profit and control were behind the wall of water that swept into the Quecreek mine. (2000)

Another convenient explanation frequently cited by managers is that the fatality was the result of a freak occurrence. A search of the Internet quickly produces large numbers of articles ascribing common workplace fatalities to freak occurrences.

> Tarrytown worker dies in freak accident: A day on the job turns deadly for a maintenance worker at the Hackley School in Tarrytown after the machinery he was using rolled over and crushed him. (Westchester News 12, 2003)

> Freak trench accident kills worker: 23-year-old man had been crushed in the collapse of a trench sidewall. (KPIX, San Francisco, CA, 2002)

> Freak accident kills man at bowling alley: A machine that resets pins appears to have come on while Devan Young, 29, was working on it. (Branch 2003)

Sometimes "freak accidents" even come in pairs:

> Williams Selyem Winery Accident: Officials are investigating the death of a 20-year-old winery worker found unconscious, apparently asphyxiated, in a production tank at Healdsburg's Williams Selyem Winery late Thursday afternoon.

> Taylor James Atkins, a Forestville resident, likely died of exposure to nitrogen gas in what would be the *second such freak accident since November 1997* [emphasis added], when . . . 49-year-old Jose Villareal was found dead at the bottom of a 50,000-gallon tank filled with nitrogen. (Calahan 1999)

As a matter of fact, the wine industry seems to be full of freak and bizarre confined-space accidents:

> *B.C. wine-country residents shocked by deaths*
> VANCOUVER (CP)—British Columbia's winemaking industry is at a loss to explain how two winemakers died in a *freak accident no one has ever heard of happening before* [emphasis added]. Victor Manola, owner of the Silver Sage Winery in the Okanagan town of Oliver, B.C., died Sunday after falling into a fermentation tank. Winemaking consultant Frank Supernak, died when he fell in, too, attempting to rescue Manola. It's believed the men suffocated because of the carbon dioxide generated in the enclosed tank by the fermenting wine. (*Vancouver Globe and Mail* 2000)

The *Merriam-Webster Dictionary* defines freak as "a seemingly capricious (unpredictable) action or event." In fact, the hazards of asphyxiation in confined spaces due to odorless nitrogen gas (as well as other asphyxiants) are well known. According to OSHA's records, at least twenty-one people died in the United States between 1990 and early 1996 in incidents involving the use of nitrogen in confined spaces. In March 1999, after investigating the death of a worker from nitrogen asphyxiation in a Union Carbide plant, the U.S. Chemical Safety and Hazard Investigation Board recommended that the National Institute for Occupational Safety and Health (NIOSH) "conduct a study concerning the appropriateness and feasibility of odorizing nitrogen in order to warn personnel of the presence of nitrogen when it is used in confined spaces" (U.S. Chemical Safety Hazard Investigation Board 1999). NIOSH has so far refused to conduct such a study.

There is clearly more work to be done on the issue of confined spaces—especially related to nitrogen asphyxiation. It may be that better guidelines or regulations should be issued. It may also be that nitrogen gas should be odorized. What is crystal clear, however, is that there is nothing freak about workers dying of nitrogen asphyxiation or other asphyxiants in confined spaces.

In her recent book, *A Job to Die For,* Lisa Cullen discusses the problem with ascribing workplace injuries and fatalities to freak accidents:

> An accident can be defined as an unexpected and unintentional happening that results in damage to people or property. Although it is common to say, "Hey, accidents happen," they are more complicated than that. In hindsight, most can be seen building from several causes, each representing a missed opportunity to step in and prevent the forthcoming damage. In fact, the safety and health profession is so averse to the term accident that the word incident has been widely substituted.

> In the workplace, few real accidents occur because the surroundings and operations are known; therefore, hazards can be identified. When harm from those hazards can be foreseen, accidents can be prevented. (2000, 5)

Unfortunately, blaming a workplace fatality on God, freak occurrences, or a careless worker is a way of thinking that the media often fall into and that some employers encourage. After all, if a workplace fatality is unpredictable and unpreventable, then no great public outcry is warranted. If someone's inattentiveness or stupidity or laziness (or drug problem) or God's will led to the death, then it is a tragedy for the family and friends, but no real investigation is needed, no lessons are to be learned, no changes in the workplace are demanded, no new OSHA regulations are needed, no enforcement is appropriate, and no wider social problems need to be addressed.

Employers often get away with blaming deaths and injuries on freak accidents and other excuses—at least in the public eye. They are typically quoted about the freak accident in a short one-day story in the local newspaper and by the time experts are found (if anyone bothers) or the OSHA report comes out explaining the employer's failure to provide a safe workplace, the local media have often lost interest.

When willful management negligence seriously injures or kills a worker, high fines or even jail may be appropriate. It is important to not lose sight of the root causes of workplace incidents while looking for a guilty party. Sometimes the press, courts, or enforcement agencies miss the real causes of incidents and misallocate the blame.

Root causes of workplace incidents often are not readily evident to untrained observers. The National Transportation Safety Board (NTSB), which investigates airplane, transportation, and pipeline accidents, and the U.S. Chemical Safety Hazard Investigation Board, which investigates incidents involving chemical manufacturers and users, look for the root causes of accidents, which usually are not found in the errors of specific individuals but in management safety systems.

Unfortunately, public blame still tends to fall on individual employees who may not be responsible for controlling a worksite's safety systems. A recent example in the *Pipeline Accident Report 1999* illustrates this point:

> On June 10, 1999, a 16-inch–diameter steel pipeline owned by Olympic Pipe Line Company ruptured and released about 237,000 gallons of gasoline into a creek that flowed through Whatcom Falls Park in Bellingham, Washington. About 1 1/2 hours after the rupture, the gasoline ignited and burned approximately 1 1/2 miles along the creek. Two 10-year-old boys and an 18-year-old young man died as a result of the accident. Eight additional injuries were documented. (NTSB 1999)

In October 2002, the NTSB issued a report identifying five main causes of the incident:

1. Damage done to the pipeline by adjacent construction five years before the accident;
2. Olympic Pipe Line Company's failure to examine the damaged pipe;
3. Olympic's failure to test safety devices at a new facility;
4. Olympic's failure to investigate and correct the repeated unintended closing of an inlet block valve; and
5. Olympic's practice of performing database development work on the computer system that controlled the pipeline while the system was being used to operate the pipeline, which led to the system failing at the same time that the pipeline break occurred.

In other words, according to the NTSB, there was a total breakdown in the entire management safety system that led to a series of interrelated system failures, ultimately concluding in a catastrophic event. Yet in December 2002, criminal indictments were handed down against three midlevel managers who happened to be operating parts of the fatally flawed system. As the father of one of the children killed insightfully noted, "These three guys are really victims of an industry. It's the industry that needs to change" (Johnson and Ho 2002).

Contrast this with a recent New York City event where there was a knowledgeable union that challenged a finding of blame against the crew supervisor. Within days of each other last fall, two New York City subway workers were struck by trains and killed while working near the tracks. The *New York Times* initially intimated that perhaps the worker was not paying attention:

> On Thursday, Nov. 21 at 11:19 a.m., Joy Antony was peering into a glass box just north of the West 96th Street subway station to test whether a warning light several yards away was working properly, when a northbound express train approached. Standing on a narrow slab of concert between tracks, he leaned back a little. Maybe the rush of air pushed him another inch. He got too close to a southbound train behind him, and something on the side caught him and scooped him under. (Wilson and Christian 2002)

Following an investigation, however, the New York City Metropolitan Transit Authority acknowledged its own failure to provide adequate staffing for flaggers while track work was being conducted. The *New York Times* reported that the "sweeping" changes to safety rules would require "hiring more workers or paying more overtime, thus costing the agency more money in a time of serious budget problems."

Yet this admission did not stop the Transit Authority from charging the supervisor, Deanroy Cox, who was on duty the night Antony was killed. The Transport Workers Union objected to the discipline, even though the supervisor was not a union member:

> Union officials said that the problem was in staffing levels and that Mr. Antony's death would have been prevented if an additional worker had been assigned to his crew. It said Mr. Cox pulled him away from his flagging assignment because he was needed to work on a subway signal.
>
> Mr. Cox "basically had to chose between flagging protection and getting his work done," said John Samuelsen, a spokesman for Local 100. "He's a symptom of a much larger problem and is now being used by management as a scapegoat." (*New York Times* 2003)

The *Columbia* space shuttle accident is, of course, perceived much differently than a typical workplace fatality. The astronauts clearly did not meet their ends because of their own inattentiveness, nor is anyone publicly dismissing it as an act of God or just a freak accident. The disaster was clearly a result of failed management safety systems (NASA 2003). People are outraged at reports of warnings unheeded and close calls unreported. We are seeing serious discussions about whether enough money is being spent to keep America's astronauts safe. Have we bumped up against technological barriers? Have we cut spending for the space program too much? Did NASA dismiss potential whistle blowers? No cost will be spared in an effort to identify the causes and develop safety systems and equipment that will make sure that nothing like that ever happens again.

Most workplace fatalities have nothing to do with stretching the limits of technology. They are caused by a failure of the employer to address well-recognized unsafe workplace conditions and implement effective safety programs in their workplaces. Saving workers' lives is much less about technological challenge than it is about better enforcement of the Occupational Safety and Health Act, more regulation of hazardous substances and unsafe working conditions, more research into occupational diseases, and more pressure on employers to take responsibility for making their workplaces safe.

Yet OSHA's budget does not approach the level needed to fulfill its congressional mandate and the NIOSH, responsible for the nation's research into workplace injury, death, and illness, is similarly underfunded. OSHA's resources remain inadequate to meet the challenge of ensuring safe working conditions for American workers:

> In FY 2003, there are at most 2,144 Federal and state OSHA inspectors responsible for enforcing the law at nearly eight million workplaces.

> At its current staffing levels and inspection levels, it would take Federal OSHA 115 years to inspect each workplace under its jurisdiction just once. In four states (Florida, Georgia, Louisiana, and Mississippi), it would take more than 150 years for OSHA to pay a single visit to each workplace. In 18 states, it would take between 100 and 149 years to visit each workplace once. Inspection frequency is better in states with OSHA approved plans, yet still far from satisfactory. In these states, it would now take the state OSHAs combined 60 years to inspect each worksite under state jurisdiction once, as compared to once every 62 years in FY 2000. (AFL-CIO 2003)

On one hand this society's failure to equip OSHA with the means to accomplish its mission is a result of the invisibility of "normal" workplace death and injury. The visibility of the space program, its place in the public imagination, and the worldwide human interest in the lives of the astronauts ensure that this society will never tolerate the same level of attention to safety in the space program that it tolerates in American workplaces.

But in another sense, the underfunding of OSHA, its lack of mission, and the dearth of media attention and public outrage over preventable workplace deaths are also the result of another, more significant failure. The labor movement and the occupational health and safety community in general have not succeeded in igniting a successful educational and political campaign within their institutions and within society that will achieve the same level of intolerance for the death and injury of American workers that we have seen for the death of the astronauts.

So, why is it important to fight for more and better media attention to workplace fatalities and injuries? In one sense, media attention is not important. No amount of media will bring back a loved one. In another sense, however, media attention that accurately reflects the causes of workplace fatalities is central to the fight for safer workplaces. Seeing the faces and knowing the stories of workers killed on the job transforms them from statistics into people—just like the people we know and love. The shock, the sorrow, and the basic unfairness and injustice of those deaths spurs action.

But media attention is not necessarily good in itself unless it accurately reflects the real causes and preventability of workplace injuries and fatalities. As long as people believe workplace injuries and fatalities are unpreventable or caused by the mistakes of individual workers, effective local or national efforts to prevent similar occurrences will be stymied.

To better educate the public and to raise the level of intolerance for workplace death, unions need to educate their members and field staff about why injuries and illnesses happen and how they can be prevented. Unions, working with health and safety professionals and activists, need to develop an active voice in media stories about workplace fatalities, emphasizing that most accidents can be prevented, identifying the safety and health standards that may have been violated, and stressing that the accident is not the fault of the worker. Unions should call for and participate in a full investigation of the incident in which their members are killed or injured, even if OSHA chooses not to investigate or if OSHA does not cover the worker. Finally, these voices must join locally and nationally on a political level to make sure that this society creates the institutions that will, in the words of the Occupational Safety and Health Act, assure "every working man and woman in the Nation safe and healthful working conditions."

More media attention may not bring back the dead, but it can result in more resources to prevent future fatalities. And more attention and more resources can safely bring home thousands of parents, wives, and husbands every year.

The nation's great emotional attachment to the lives and deaths of the astronauts may be human nature: our human and historical admiration for mankind's explorers. And rightly so, but while we are mourning the astronauts, let us not dismiss the tragedies that befall thousands of "regular people" every year and that, collectively, we can take actions to prevent.

References

"Achievements in Public Health, 1900–1999: Improvement in Workplace Safety—United States, 1900–1999." 1999. *Morbidity and Mortality Weekly Report* 48, no. 22 (June 11): 461.

AFL-CIO. 2003. *Death on the Job: The Toll of Neglect—A National and State-by-State Profile of Worker Safety and Health in the United States.* 12th ed. Washington, DC: AFL-CIO.

Barab, Jordan. 1997. "Union Perspective on Workplace Violence." *Arbitration 1996, At the Crossroads.* Washington, DC: Bureau of National Affairs, 1997.

Branch, Alex. 2003. "Freak Accident Kills Man at Bowling Alley." *Wichita Eagle,* January 9.

Brown, David. 2003. "U.S. Troops' Injuries in Iraq Showed Body Armor's Value." *Washington Post,* May 4.

Bush, George W. 2003. "Remarks by the President at the Memorial Service in Honor of the STS-107 Crew, Space Shuttle Columbia," February 4.

Calahan, Mary. 1999. "Williams Selyem Winery Accident." *Santa Rosa Press Democrat,* January 9, published in Vinginuity. www.eresonant.com/pages/risk/risk-tankdeath1.html.

Conway, Mike. 2003. "Charge Stand in Dairy Deaths." *Modesto Bee,* May 3.

Cullen. Lisa. 2002. *A Job To Die For: Why So Many Americans Are Killed, Injured or Made Ill at Work and What to Do About It.* Monroe, ME: Common Courage Press.

Ferreira, Alcides. 2003. "Labour Day Thought: Paycheques Not Worth Dying For." *InfoBrazil.com,* April 26.

Horrell, Steve. 2003. "Fatal Fall at Fairmount Accidental." *Edwardsville Intelligencer,* April 25.

Johnson, Tracy, and Vanessa Ho. 2002. "In Deal, Olympic Pipe Line, 3 Workers Admit Guilt in Blast." *Seattle Post-Intelligencer,* December 12.

Kniesner, Thomas J., and John D. Leeth. 2000. "Abolishing OSHA." *Cato Review of Business & Government,* January. www.cato.org.

Liberty Mutual Research Institute for Safety. 2003. *Workplace Safety Index 2003.* www.libertymutual.com.

Loeb, Paul Rogat. 2003. "Those We Mourn and Those We Ignore." *Seattle Times,* February 19 and May 2. www.infobrazil.com/conteudo/front%5Fpage/analysis/af4185.htm.

Mayo, John. 2003. "Time's Right to Get 'MSDD' About Workplace Deaths." *Houston Chronicle,* February 17.

McColester, Charles. 2002. "Less Than Miraculous: The Near-Disaster at Quecreek Mine." *The Nation,* March 16. www.hhs.iup.edu/laborcenter/ProgramsProjects/articles/index.shtml.

National Aeronautics and Space Administration (NASA). 2003. *Report of Columbia Accident Investigation Board,* August 26. www.nasa.gov/columbia/home/CAIB_V011.html.

National Transportation Safety Board (NTSB). 1999. "Pipeline Rupture and Subsequent Fire in Bellingham, Washington," in *Pipeline Accident Report,* Number PAR-02/02, June 10. www.ntsb.gov/publictn/2002/PAR0202.pdf.

New York Times. 2003. "Transit Inquiry Faults Supervisor in Subway Worker's Death," April 10.

Occupational Safety and Health Administration. 1998. *Guidelines for Preventing Workplace Violence for Health Care and Social Service Workers* (revised: 1998); and *Recommendations for Workplace Violence Prevention Programs in Late-Night Retail Establishments.* Washington, DC: U.S. Department of Labor, OSHA.

Sysco Food Services. 2000. Memo from Sysco Food Services of Baltimore to All Delivery and Warehouse Associates, January 12.

U.S. Bureau of Labor Statistics (BLS). 2001. *National Census of Fatal Occupational Injuries Data.*

U.S. Chemical Safety Hazard Investigation Board. 1999. *Nitrogen Asphyxiation.* Report No. 98–05-I-LA. February 23.

Vancouver Globe and Mail. 2000. "B.C. Wine Industry Shocked by Freak Deaths," November 11.

Westchester "News 12." 2003. April 25.

Wilson, Michael, and Christian, Nichole. 2002. "Killed in Subway They Tried to Make Safer for Others." *New York Times,* November 29.

2

Criminal Neglect

How Dangerous Employers Stay Safe From Prosecution

Rory O'Neill

Hundreds of thousands die at work each year, many as a result of clear negligence and corner-cutting on the part of their employers. But where these crimes would usually result in a prison sentence when committed outside the workplace, behind the factory doors employers can almost always kill with impunity. Unions worldwide are now calling for corporate accountability to extend to workplace safety crimes and for jail sentences for the most dangerous employers.

When in late 2001, following the death of Emilio Palomares, prosecutors filed what was believed to be California's first involuntary manslaughter case (in the death of a farm worker), the prosecution had high hopes of success. The case against Donald William Beeman, a grower who faced four years in prison and $650,000 in fines if convicted, was "quite rare, but it's going to be more and more common," said deputy district attorney Kyle Hedum. A new California statute, AB 1127, had elevated to a potential felony any serious Labor Code violation that resulted in a worker's death or serious injury. The case was the first to be tried under the new system. Palomares's leg was severed while attempting to remove clogged corn from a harvester operated by Beeman, his boss. He died at a local hospital due to massive loss of blood. Despite the new legal code, however, a Yolo County judge ruled on May 17, 2002, that there was insufficient evidence to take the involuntary manslaughter case to trial ("AB 1127 Test Case" 2002).

That a worker died in what was evidently a needless and preventable accident and no one was found guilty of any crime is not what is extraordinary. What makes this case stand out is that manslaughter charges were brought in the first place. Employers in the United States are occasionally jailed for safety crimes, but these cases are memorable as much for their rarity as for extraordinary abuse deemed necessary to merit a jail sentence.

In October 2002 the U.S. Supreme Court ruled that the conviction of an Idaho fertilizer company owner and his record seventeen-year prison sentence should stand. Allan Elias was jailed after an incident that left employee Scott Dominguez, twenty years old at the time, with permanent brain damage. The owner of Evergreen Resources, located near Soda Springs, Idaho, had ordered four employees to clean out a storage tank containing one to two tons of hardened cyanide waste ("US Top Court Rejects Appeal" 2002).

A decade before, on September 3, 1991, a fire at Imperial Foods Products, a chicken processing plant in Hamlet, North Carolina, killed twenty-five and injured over fifty others. Some exits were blocked; others were intentionally locked. There was no sprinkler system and no evacuation procedure. Although the plant was visited daily by Department of Agriculture inspectors to ensure food safety, the workplace had not had a single visit in its eleven-year history from the workplace safety watchdog, North Carolina Occupational Safety and Health Administration. The plant was fined $808,000. Factory owner Emmett Roe was convicted on twenty-five counts of involuntary manslaughter. He served just over four years of a nineteen-year sentence. Back in Hamlet, Patricia Hatcher, seriously injured in the fire, became its twenty-sixth victim when she succumbed to her injuries in 2000.

In January 2004, New York State Supreme Court Justice Rena Uviller jailed construction boss Philip Minucci, thirty-two, after five immigrant workers were killed in what was called a "tragic certainty" rather than an accident. Minucci received a three- to ten-year sentence. The five laborers killed were among twenty masonry workers on a job in Manhattan on October 24, 2001. The majority of the workers were illegal immigrants paid $7 an hour in cash. Minucci, owner of Tri-State Scaffolding & Equipment Supplies, designed and built the 130-foot scaffold despite not being a licensed architect or engineer. The judge said the case illustrated how "astonishingly ineffectual" the federal government's Occupational Safety and Health Administration (OSHA) has been in protecting workers' lives ("Cavalier Boss Gets Jail Term" 2004).

The jail terms served by Roe, Elias, and Minucci do not serve as much of a warning to would-be dangerous employers. Foreseeable, preventable deaths and grievous injuries at work rarely result in court action, let alone jail time. Addressing a May 1, 2003 meeting of the House Appropriations Subcommittee on Workforce Protections, Representative Rosa DeLauro

(D-CT) appealed to OSHA administrator John Henshaw for support of tougher criminal sanctions. She said: "From 1972 to 2001, there have been at least 200,000 on-the-job deaths, 151 referrals for criminal investigation, and 8 cases resulting in jail time" ("Rep. DeLauro Presses OSHA" 2003).

DeLauro, like many across the United States, had been horrified by a January 2003 series in the *New York Times* that described how since 1995 nine workers died in a single company—McWane, Inc., a Birmingham, Alabama, conglomerate. In other settings this would be described as the work of serial killer. In the workplace, where eighteen U.S. workers die every day, it is a "misdemeanor."

The year ended with federal workplace safety agency OSHA facing a storm of criticism over its lax approach to safety enforcement. A December 2003 series of award-winning investigative reports in the *New York Times* revealed that over the past twenty years OSHA has investigated more than 1,200 deaths caused by "willful" safety violations of the employer—cases where the employer knowingly sends a worker into a dangerous situation—but did not seek prosecution in more than 90 percent of those cases ("Employers Getting Away" 2004).

Fine Times for the Bosses

The situation in the United States is shocking, but not unique. A U.S. worker runs the risks of dying at work about four times that of a worker in the UK. Average fines are marginally higher in the United States, but employers are less likely to face manslaughter charges. As of April 2003, the UK has seen nine work-related manslaughter cases, resulting in the conviction of five companies, eight directors, and two business owners, as recorded in the Centre for Corporate Accountability's work manslaughter database (CCA 2003).

While average fines are increasing in the UK and while there is at least a remote prospect of a death resulting in manslaughter charges, the trend in the U.S. is not so encouraging. Peg Seminario, director of occupational safety for the U.S. union federation AFL-CIO, warned in 2003 that the average fine for a willful safety violation had decreased under the George W. Bush administration by 26 percent to $26,888 in 2002—and maximum fines have not been increased since 1991, the year of the Hamlet fire.

Britain's health and safety law, the Health and Safety at Work Etc. Act of 1974, is criminal law, and offenders are legally considered to be criminals—although one would have trouble determining this from the penalties imposed. Safety fines for 2001–2 totaled £10,901,263 (US$17.8 million) from almost 900 cases, at an average of £12,194 (US$20,000) per case. The average fine for work-related deaths is £20,000 to £30,000 (US$33,000–$50,000).

Average fines in the two most deadly sectors, construction, less than $12,000, and agriculture, less than $3,400, remained the lowest for any sector (HSE 2002a). Researchers at Income Data Services found that chief executive officers at the UK's top 100 companies saw their average earnings pass the £1.5 million (US$2.8 million) mark in 2002 ("Deadly Business" 2002). Wherever one looks, there is little evidence of this sort of corporate generosity extending to those affected by corporate neglect. Neither is there much evidence of justice.

In South Africa, British multinational Cape PLC, which reneged on an original £21 million (US$39 million) asbestos disease compensation deal agreed on in 2001, in 2003 agreed to pay £7.5 million (US$13.9 million) to 7,500 claimants. Aditi Sharma of Action for Southern Africa said 200 severely ill asbestos victims died while Cape stalled. No Cape executive in the UK or South Africa has faced charges relating to the deaths ("Asbestos Deal" 2003).

In Japan, six executives who allowed workers to use buckets to fill a tank with uranium, resulting in two deaths and widespread radiation contamination in the country's worst nuclear accident, escaped jail in 2003. All six received suspended prison terms, with one, the former head of the Tokaimura plant where the accident occurred, also fined. All six admitted negligence in the 1999 incident. The court also imposed a fine of ¥1 million (US$8,700) on plant operator JCO Company ("No Jail" 2003).

In Thailand, court proceedings concluded this year on cases relating to the May 10, 1993, Kader factory fire that killed 188 and injured 460 employees. A union working group is monitoring the case after the Nakhon Pathom Provincial Court in March 2003 sentenced one defendant—a worker—to ten years in prison and fined the factory ฿520,000 (US$12,500). The court acquitted fourteen other defendants, including all the managers, a director, and a shareholder who had faced charges. An earlier investigation had found the poor state of the factory, which had no fire escapes, had contributed to the deaths ("The 10th Anniversary of the Kader Fire" 2003).

In Italy, charges of "massacre" against managers of a polyvinyl chloride (PVC) factory where unions exposed an epidemic of work-related cancers, failed in the courts in November 2001. Twenty-eight senior managers of petrochemical company Montedison (subsequently Enichem) were acquitted by an Italian court, despite evidence of a conspiracy to cover up evidence of health effects, including angiosarcoma.

Treated Like Animals?

It is not that safety authorities and governments worldwide are particularly hard on corporate crime, but that they are especially soft on corporate safety crime. Nowhere is neglect and harm treated with the same casual disregard as it is at work, wherever one works.

On the same day that the UK safety enforcement agency, Health and Safety Executive (HSE)—the equivalent of OSHA—released its 2001–2 report on fines, Birmingham City soccer supporter Michael Harper was arrested for invading the pitch (i.e., the playing field) to taunt opposition players during a match. Although no one was injured, he was jailed for four months and banned by Birmingham Magistrates from attending any designated football matches for the next six years. Simon Vallor, twenty-two, was jailed for two years in January 2003 for unleashing a computer virus. In the same month, Graham Ellison, forty-four, was sentenced to four months for having an untidy garden. The factory cat has a greater chance for justice than the factory worker. The UK's Royal Society for Prevention of Cruelty to Animals (RSPCA) reported in April 2003 that offenses involving violence toward animals have risen with a total of fifty-seven prison sentences imposed in 2002, up from forty-six in 2001. Only one employer received a jail sentence for a workplace safety crime in this two-year period—and that was overturned on appeal and replaced with a £1 fine.

U.S. author and professor Jeffrey Reiman, writing in *The Rich Get Richer and the Poor Get Prison,* commented on this legal double standard:

> Is a person who kills another in a bar brawl a greater threat to society than a business executive who refuses to cut into his profits to make his plant a safe place to work? By any measure of death and suffering the latter is by far a greater danger than the former. Because he wishes his workers no harm, because he is only indirectly responsible for death and disability while pursuing legitimate economic goals, his acts are not called "crimes." Once we free our imagination from the blinders of the one-on-one model of crime, can there be any doubt that the criminal justice system does not protect us from the gravest threats to life and limb? (2001, 85)

This legal double standard works everywhere. Make a false claim for workers' compensation and you can expect jail time, whether you live in the United States or Australia. Break the law and cause someone to require workers' compensation, you go free.

A World of Difference

Work is a significant contributor to global mortality and morbidity. International Labor Office figures released in 2003 report 2 million people are killed worldwide by their work each year. For each of the 355,000 fatal accidents, there are some 500 to 2,000 other, nonfatal injuries. Add in work-related cancer deaths (32 percent of the annual workplace death toll), circulatory diseases (23 percent), and communicable diseases (17 percent), and work emerges as a major killer ("Two Million Killed" 2003).

If only the smallest proportion of this harm was the result of management negligence—and the evidence suggests most deaths are preventable and many would have been prevented if the employer had met minimum legal duties—then tens of thousands of employers each year could be culpable, and tens of thousands of employers should be facing jail time. Corporate accountability in other areas has never had more attention—rightly so, as entire workforces have seen corporate pirates plunder their pensions and sink their jobs.

But there is a growing recognition that the safety of the company finances should not be the only consideration when policing corporate behavior. The July 3, 2002, issue of *Focus on the Corporation,* commenting on U.S. government proposals to act against corporate financial abuses, reported: "Given what is now the apparent blatant corporate disregard for the law, even in areas where executives are most closely watched, what should we expect is occurring elsewhere? What's happening with consumer rip-offs, sales of unsafe products, endangerment of workers, pollution of the environment?" The article concludes: "Cracking down on corporate crime—the mantra of the moment—cannot be limited just to financial crime, already the most policed form of corporate wrongdoing" ("Cracking Down on Corporate Crime—Really" 2002).

Six months earlier, the January 26, 2002, edition called for an end to "unaccountable accounting," and the creation of a corporate accountability commission that would respect the accountants' dictum "if you don't count it, it doesn't count" and would consider "intangibles and externalities" as real occupational injury costs. "For example, when workers were injured, you would not merely report the cost of the in-house nurse and the insurance, but you would also report the cost to the worker of loss of the leg, offset by any benefits you might provide to the worker," the article read ("Unaccountable Accounting" 2002). In the United States, though, the reality is that accountability for safety crimes is being eroded by explicit legal maneuvers and by stealth.

Carrots and Sticks

Immediately below the logo on the OSHA Web site, www.osha.org, and adorning every national OSHA news release, is the legend: "Safety & Health Add Value To Your Business. To Your Workplace. To Your Life." Your life may have value, but it seems that value is lower than it used to be. While safety fines have dropped under the Bush administration, voluntary "cooperative programs," "alliances," and "strategic partnerships" have flourished.

OSHA's ergonomics standard, introduced in the last days of the Clinton administration, was cut as one of the first acts of the Bush administration (AFL-CIO 2001a; Mogensen 2003). The result was a near total cessation of ergonomics enforcement. OSHA had to go back to using the General Duty Clause of the Occupational Safety and Health Act to catch the worst offenders. In March 2003 it proposed its first employer penalties for repetitive strain injury risks, after a ten-month enforcement-free period. The unlucky recipients of OSHA enforcement include nursing home company Alpha Health Services, Inc., which initially faced three $900 penalties for ergonomics violations. Metal fabricating company Security Metal Products Corp. faced a proposed penalty of $5,600 for a material handling citation. The companies agreed not to challenge the penalties—in exchange for which OSHA halved the fines ("It's Enforcement, But Not as We Knew It" 2003).

On December 27, 2001, the White House revoked regulations that would have barred companies that repeatedly violate environmental and workplace standards from receiving government contracts. Laurence E. Gold, a counsel for AFL-CIO, commented: "If somebody is engaged in repeated pervasive violations of the law, that company should not be receiving a government contract." A statement from AFL-CIO noted: "In one year, according to the U.S. General Accounting Office, 261 federal contractors with 5,121 violations of health and safety laws among them received $38 billion in federal contracts. Breaking U.S. labor laws didn't prevent 80 other firms from getting $23 billion in taxpayer-financed projects" (AFL-CIO 2001b).

Safety Pays?

The message was clear enough. Safety is not a priority and self-regulation is the only regulation one is likely to get. Advocates of this approach say safety is in everybody's interest—it is not a matter for litigation, but a matter for cooperation. And, as an added incentive, "safety pays."

The argument sounds fine but does not bear even the most superficial scrutiny. *The Role of Managerial Leadership in Determining Workplace Safety Outcomes,* an international research review commissioned by UK's HSE and published in 2002, concluded that approaches based primarily on the enlightened self-interest of employers and managers—the "safety pays" argument—are seriously flawed, and cites international studies to support its case (HSE 2002b).

A study that investigated the impact of workplace catastrophes on shareholder value in fifteen major corporations found that "in all cases the catastrophe had a significant negative initial impact on shareholder value. However, in some cases, the net impact on shareholder value (after fifty trading days) was actually positive. . . . Such evidence illustrates that organizations often do not suffer financially, even from serious incidents." In the case of U.S. multinational Union Carbide—now Dow—and its Bhopal, India, pesticides plant, the report says: "While lower level managers may have suffered financially and lost their jobs, for senior managers, whose decisions are crucial to safety, their inattention to safety paid off handsomely." A culpable homicide charge against the company's former CEO, Long Island resident Warren Anderson, still stands in India, where Anderson is classed as "an absconder." The 1984 explosion has been linked to 16,000 deaths ("Has the World Forgotten Bhopal?" 2000).

The HSE research report (2002b) adds that in the multinational BHP mining empire Moura Mine disaster in Australia where four miners drowned in an in-rush of water from an adjacent shaft, "the catastrophe did not have any serious financial impact on the most senior executives of the company. Again the parent company did not suffer financially, and again, senior managers were shielded from any untoward financial consequences." The report concludes that "while the moral case of investment in health and safety is indisputable, the argument that safety pays is spurious."

It recommends a different sort of strategy to deal with errant employers. The report says corporate manslaughter legislation "should act as a powerful deterrent to help prevent needless injuries and deaths whilst at the same time punishing the grossly negligent," adding that effective enforcement "is likely to be strengthened further" by a corporate killing law (HSE 2002b).

The Case for Corporate Killing Laws

Arguments for greater accountability for workplace safety crimes are not new. Each new workplace disaster prompts calls for the guilty to be brought to justice. In Canada, a 1992 mine disaster became the focus of a union campaign for industrial manslaughter laws. On May 9, 1992, when coal dust and methane gas exploded in the southwest section of the underground Westray coal mine in Plymouth, Nova Scotia, the immediate effect was a devastating fire, a blast that ripped the roof off the mine entrance, and the death of twenty-six miners.

In Australia, the explosion at Exxon's Longford chemical plant in 1998 spurred calls for state and national laws. A commission found in September 2001 that Exxon was responsible for the two deaths in what was described at an earlier sentencing hearing as a "grievous, foreseeable and avoidable" explosion, adding new impetus to the campaign ("Esso Has Been Found Guilty"

2001). In the UK, a sequence of disasters, from the 1988 Piper Alpha oil rig fire that claimed 167 lives to a spate of recent rail crashes, added to a growing grassroots movement, bringing together union and safety activists and bereaved families. The disasters come apace—the 2000 Chowdhury Knitwear factory fire in Bangladesh that killed forty-eight (ITGLWF 2002); the 1993 Kader factory fire in Thailand with 188 dead ("The 10th Anniversary of the Kader Fire" 2003); and the African Rubber Products factory fire in Nigeria in 2002, now thought to have killed thirty-seven ("Nigeria" 2003)—and all lead to renewed calls for justice for those killed and for the families left behind.

In 2003, for the first time, the international trade union movement made its first united call for greater accountability of employers for workplace safety crimes. This was the theme of the worldwide International Workers' Memorial Day, April 28, 2003, coordinated globally by the International Confederation of Free Trade Unions (ICFTU), the sectoral global union federations, and national union centers ("International Workers' Memorial Day" 2003).

According to David Bergman, director of the London-based Centre for Corporate Accountability, the case for pursuing dangerous employers is clear. "Company directors are the individuals within a company with the most power to determine whether or not the company conducts its activities safely or not. They control the company's resources and make key decisions about how the company will run." He added: "It is so important that company directors take their responsibilities seriously, that the threat of imprisonment should be one of the sentences that courts have the power to impose when a director is convicted" (Centre for Corporate Accountability 2003).

Canadian law professor Harry Glasbeek of York University in Toronto, Ontario, also believes company directors and their shareholders should be called to account where safety breaches cause harm to workers. In a 2001 speech delivered in Victoria, Australia, he said: "The employer has limited risks, certainly no physical ones and the costs are usually passed onto workers and consumers. So the people who take the most amount of risk have the least to say. . . . I would suggest we go as far as charging the shareholders of grossly negligent companies. Just as you would charge someone who benefited from the criminal behavior of a thief by receiving stolen goods."

The effect of an industrial manslaughter law goes beyond the penalty imposed, he said. "Under industrial manslaughter legislation a grossly negligent employer would be denounced and stigmatized if a worker was killed. It would be an important step to have industrial manslaughter legislation even though it would lead to very few convictions" ("Professor Harry Glasbeek" 2003).

The Laws Are Coming

On May 20, 2003, the UK government announced its tentative timetable for the introduction of a corporate killing law. The move was prompted by a series of recent parliamentary maneuvers backing a highly organized and high profile union campaign for a new law. Announcing the plans, Home Secretary David Blunkett said: "There is great public concern at the criminal law's lack of success in convicting companies of manslaughter where a death has occurred due to gross negligence by the organization as a whole. The law needs to be clear and effective in order to secure public confidence and must bite properly on large corporations whose failure to set or maintain standards causes a death" (Home Office 2003).

The proposals, however, fall well short of the demands of campaigners. The UK government says it will bring corporations to justice. But corporations do not serve jail time, and unions, victims' organizations, legal groups, and even the Institute of Directors—the industry umbrella group for the top company directors in the UK—say the provision will be severely weakened unless directors are made to consider their liberty alongside their safety priorities. A June 7, 2003, editorial in the medical journal *The Lancet* concluded: "The UK government's new law may well be a step towards making companies more financially accountable for their actions, but until chief executives are made directly responsible for decisions that lead to injury, it is unlikely that the huge toll of work-related injuries will fall" ("Who Will Take Responsibility" 2003).

In June 2002, on the tenth anniversary of the Westray mine disaster in Nova Scotia, the Canadian Labour Congress (CLC) renewed its call in support of legislation to amend the Criminal Code to hold negligent employers responsible for workers killed on the job. In a presentation to the House of Commons Standing Committee on Justice and Human Rights, Hassan Yussuff, CLC executive vice president, said: "Effective legislation would result in prosecution of cases where there is reason to believe that a corporation or its managers were guilty putting workers' lives at risk. It will also serve as a warning to directors and officers of a corporation that avoiding responsibility is itself a crime." The corporation that owned Westray "and its directors should have been held criminally responsible. . . . To date, not a single party or individual has been brought to justice over the entirely avoidable Westray disaster. It is episodes such as Westray and others where a worker has been culpably killed or injured that amendments to the Criminal Code should address" (CLC 2002).

The Australian Council of Trade Unions (ACTU) has called for similar legislation. Draft laws have already been proposed in the state parliaments in Victoria, New South Wales, and in Queensland. The union campaign was given impetus by the findings last year that Exxon was responsible for two deaths in a "grievous, foreseeable and avoidable" 1998 explosion at its Longford, Australia, plant ("Longford" 2001).

Progress in Australia has been faltering, however. In Victoria a preelection promise to back a statewide industrial manslaughter law was dropped when the government won reelection. Had it been introduced, the Crimes (Workplace Deaths and Serious Injuries) bill, tabled in the Victorian parliament in November 2001, would have meant companies whose gross negligence caused the death of a worker would have faced fines up to $5 million Australian (US$3.28 million). Senior officers of those companies face five years in jail or hefty fines. Companies that cause serious injury through gross negligence could be fined over $1 million Australian and senior managers could be personally fined $65,000 or jailed for two years.

In New South Wales the state government introduced measures short of new law in a bid to head off critics—measures included the creation of a specialist unit within WorkCover, its safety enforcement agency, to investigate every workplace fatality with a view to criminal prosecutions. Unions in the state have continued their campaign, however. In January 2003 massive billboards calling for the introduction of industrial manslaughter legislation were booked at locations around Sydney, the New South Wales state capital. The billboards feature the body of "A Worker," with a toe tag stating the cause of death as poor safety practices. The billboards are backed by the Construction, Forestry, Mining and Energy Union (CFMEU) and the Australian Manufacturing Workers' Union (AMWU) in New South Wales.

Peter McClelland of CFMEU said safety crimes must be taken seriously, with "a range of options including prison and disqualification from being a company director." Wayne Phillips of AMWU said company directors should be charged with industrial manslaughter if they were aware of a safety problem, did not act, and a worker died as a result ("Deadly Business" 2002).

The demands made by unions are broadly similar wherever they are proposed: For example, in New South Wales legal changes proposed in July 2002 included:

- New provisions to hold corporations and/or senior officers liable for a charge of industrial manslaughter where the decision (or omission of a decision) by the boardroom or senior management results in the death of an employee. The legislation would allow the courts to look at the conduct of a corporation and its officers as a whole, rather than the currently inadequate common law restrictions.

- A range of proposed sentencing options included, but were not limited to: imposing fines; disqualifying offenders from holding positions of corporate respon-sibility such as directorships, senior officership, and so on; ordering payments of compensation to the victim's dependents; disqualifying offenders from holding government contracts; and enforcing terms of imprisonment.

In Britain, the Trades Union Congress, the UK's equivalent of the AFL-CIO, has called for a series of measures, including:

- tougher enforcement of health and safety laws and higher fines;
- corporate killing legislation, and the extension of the possibility of a jail sentence for all breaches of health and safety law;
- more powers for union safety reps to influence how health and safety are managed at their workplace; and
- more inspectors to enforce the law. (TUC 2003)

The Laws Are Here

Campaigns for justice after workplace safety crimes recently had notable successes. Two new corporate killing laws, a national law in Canada and a state law in Australia, reached the statute books in 2003 and are now in effect.

Unions in Canada say the new corporate safety crimes law will at last make employers accountable for dangerous workplace practices. Bill C-45, an Act to Amend the Criminal Code (Liability of Organizations), was dubbed "the Westray bill," because it followed the lengthy campaign by the surviving families, former workers, and the Canadian United Steelworkers of America union at the Westray mine in Stellarton, Nova Scotia. The Canadian law makes it easier to prosecute companies, public bodies, and other "unincorporated associations" following a workplace death, but also where there has been a serious injury. Occupational diseases are also within its scope as it places a duty on company directors and others with supervisory responsibilities "to take reasonable steps to prevent bodily harm to that person, or any other person, arising from that work or task" ("International Reforms" 2004).

On March 1, 2004, a new industrial manslaughter law took effect in the Australian Commonwealth Territory (ACT), including Canberra, the home of the business friendly national government. The move, which has been welcomed by unions, came despite the strong opposition of business and the national government. The Crimes (Industrial Manslaughter) Amendment Bill 2002 is the first law of its type in Australia. ACT industrial relations minister Katy Gallagher said: "The government's position is clear—if a workplace death occurs and that death can be attributed to the employer, then the death should be treated with the seriousness it deserves." She added: "This legislation simply ensures that companies can be held responsible where their criminally reckless or negligent conduct causes the death of a worker" ("International Reforms" 2004).

The ACT law imposes new duties on "senior officers" of companies and allows for their prosecution if they negligently or recklessly cause the death of a worker. The law allows for a maximum penalty of twenty-five years in prison on conviction.

Developments in the United States

In the run up to the new millennium, union-commissioned researchers ranked organized labor's top achievements of the twentieth century. The introduction of the Occupational Health and Safety Act in 1970 made the Top 10, and was the most recent union highlight. The safety law, signed by Republican president Richard M. Nixon, came after almost a century of struggle (AFSCME 1999).

A generation later, underenforcement of the law and erosion of traditional workplace protections—particularly the decline in permanent union jobs and the rise of nonunion, insecure, service sector McJobs—has undermined the effectiveness of this law. High profile cases like McWane, Inc., have helped focus attention on just how easily corporations can kill and maim and still evade real justice.

But the Bush administration, despite it deregulatory credentials and its business friendly bent, is not immune to pressure. In the wake of the McWane media coverage the administration announced a new enforcement policy to give OSHA more power to crack down on companies that persistently flout workplace safety rules. The new approach will mean employers could face automatic inspections at all their worksites when an egregious safety violation, for example a fatality, occurs at one.

There are early signs, however, that this measure is not enough to satisfy critics. Senator Jon Corzine (D-NJ) circulated a letter to other senators in February 2003 seeking support for a proposed Wrongful Death Accountability Act, which would increase to ten years from six months the maximum criminal penalty for employers who cause the death of a worker by willfully violating safety laws. The senator said that a "staggering" 6,000 workers were killed every year and a further 50,000 die of work-related illness (Corzine 2003a). At a June 9, 2003, press conference he said: "Causing the death of an employee on the job through willful violation of worker safety standards should not be treated as a trivial federal offense." He added, "Because the penalty upon conviction is so minimal and the resources required to bring a case to trial are substantial, federal prosecutors have been reluctant to prosecute flagrant safety violations that cause death. As a result, the deterrent value of the criminal statute has eroded significantly." It was also unfair to companies that take their safety responsibilities seriously, he said. "Allowing violators to get away with breaking the law not only puts workers at risk, it creates an unfair competitive advantage and penalizes businesses that do maintain a safe workplace" (Corzine 2003b).

In May 2003, opponents of corporate crime in California took the first steps toward a "three-strikes-and-you're-out" law for businesses guilty of illegal practices, including labor offenses. "If this is good enough for individual felons in California, it's certainly appropriate for the Enron's of the world," said Carmen Balber, a consumer advocate for the Foundation for Taxpayer and Consumer Rights. Together with other consumer groups and organizations representing environmentalists, labor unions, and trial lawyers, the foundation is backing a bill that would bar a corporation from doing business in California if it is convicted of three felonies in a ten-year period. The bill, which would cover illegal financial dealings, consumer and environmental protection, civil rights, union rights, and employment laws, passed its first legislative hurdle in May 2003. Balber said the law would act mainly as a deterrent that would result in few third strikes. "Fines sometimes are not enough to prevent corporate crimes but public embarrassment and banishment would," she said ("USA: Three Strikes" 2003).

The Union Effect

In 2003, on the ninety-second anniversary of New York's worst industrial disaster, the Triangle Shirtwaist Factory fire, the site of the fire was designated an official historic monument. The disaster on March 25, 1911, in which 146 young women trapped in the burning sweatshop were killed, spurred a union movement among garment workers and led to the introduction of safety and antisweatshop laws.

Around 6,000 workers still die each year in U.S. workplace accidents—and the Imperial Foods, McWanes, and Bhopals show that employers in the United States and around the world can on occasion show the same cavalier disregard for workplace health as their industrial predecessors. Herbert Abrams's "Short History of Occupational Health" concludes that organized labor has been central to most workplace health and safety improvements, from the industrial revolution to the present (Abrams 2001).

There still is clearly a job to be done. Nearly a century on from the Triangle fire, unions, spurred on by a desire to end needless deaths, are campaigning for new laws and new protections. Today, unionized workplaces are safer workplaces and are more likely to be inspected by safety enforcement agencies. And active, organized workplaces are the safest workplaces (Ochsner and Greenberg 1998). Despite this, eighteen workers in the United States still die each day while doing their job and employers continue to kill with virtual impunity. Killing people at work requires criminal neglect. It is something neither unions nor law enforcers should allow to continue.

References

"AB 1127 Test Case Fails at Hearing." 2002. *CalOSHA Reporter,* May.

Abrams, Herbert K. 2001. "A Short History of Occupational Health." *Journal of Public Health Policy* 22, no. 1: 34–80.

AFL-CIO. 2001a. "Statement by AFL-CIO President John Sweeney on the House Vote to Repeal the Ergonomics Standard." AFL-CIO press release, March 7.

———. 2001b. "AFL-CIO Vigorously Opposes Elimination of Contractor Responsibility Rules." AFL-CIO press release, December 27.

AFSCME. 1999. "20th Century Triumphs of American Workers." Remarks by AFSCME President Gerald W. McEntee, September 1, 1999.

"Asbestos Deal Won't Bring Back the Dead." 2003. *Hazards,* no. 82 (April–June): 22–27.

"'Cavalier' Boss Gets Jail Term for Scaffold Deaths." 2004. Trades Union Congress. *Risks* 140, January 24.

Centre for Corporate Accountability (CCA). 2003. Work Manslaughter Online Database. www.corporateaccountability.org/manslaughter.htm.

Canadian Labour Congress (CLC). 2002. "Remembering Westray Workers." Press release, June 19. www.action.web.ca/home/clchealt/en_readingroom.shtml?sh_itm=1f265d0e39fd0160db05542c6d583809&AA _EX_Session=c650e2e4bee9dc2e5d8962bc28d99e1 4.

Corzine, Jon S. 2003a. "Corzine on Worker Safety." Press release of Senator Corzine, April 29. www.senate.gov/~corzine/press_office/record.cfm?id =203219.

———. 2003b. "Corzine Calls for Severe Punishment of Employers Who Violate Safety Laws." Press release of Senator Corzine, June 9. Available at: www.senate.gov/~corzine/press_office/record.cfm?id=204776.

"Cracking Down on Corporate Crime, Really." 2002. *Focus on the Corporation,* July 3. Available at: www.lists.essential.org/pipermail/corp-focus/2002/000120.html.

"Deadly Business: Unions Target Workplace Safety Criminals." 2002. *Hazards,* no. 80 (October–December): 4–5. Available at: www.hazards.org/deadlybusiness.

"Employers Getting Away with 'Willful' Deadly Violations." 2004. Trades Union Congress. *Risks* 138, January 10.

"Esso Has Been Found Guilty of 11 Breaches of the OHS Act and Fined $2 million." 2001. Victorian Trades Hall Council, August 6. Available at: www.vthc.org.au/campaigns/20010806_Esso.html.

"Has the World Forgotten Bhopal?" 2000. Editorial. *Lancet* 356, no. 9245: 1863.

Health and Safety Executive (HSE). United Kingdom. 2002a. *Offences and Penalties Report for 2001/02*. UK, November. HSE enforcement action Web page, www.hse.gov.uk/enforce

———. 2002b. *The Role of Managerial Leadership in Determining Workplace Safety Outcomes*. HSE research report, RR044. Available at: www.hse.gov.uk/research/rrhtm/rr044.htm.

Home Office. 2003. "Government to Tighten Laws on Corporate Killing." Home Office news release, May 20.

"International Reforms Eclipse British Proposals." 2004. *Corporate Crime Update*, no. 8 (Winter). Centre for Corporate Accountability.

"International Workers' Memorial Day, 28 April 2003." 2003. Available at: www.hazards.org/wmd.

ITGLWF. 2002. "Global Union Calls for Release of Chowdhury Campaigner." International Textile, Garment and Leather Workers' Federation news release, January 9. Available at: www.itglwf.org/displaydocument.asp?DocType=Press&Index=331&Langu age=EN.

"It's Enforcement, But Not as We Knew It." 2003. *Hazards*, no. 82 (April–June): 12.

"Longford Was Grievous, Tragic and Avoidable." 2001. Victorian Trades Hall Council news release, August 6. Available at: www.vthc.org.au/media/general_news/20010806_Esso.html.

Mogensen, Vernon. 2003. "The Rise and Fall of OSHA's Ergonomics Standard." *WorkingUSA* 7, no. 2 (Fall): 54–75.

"Nigeria: Union Health and Safety Plan Revealed." 2003. Trades Union Congress. *Risks*, no. 89, January 18.

"No Jail After Japanese Nuke Plant Deaths." 2003. *Hazards*, no. 82 (April–June): 12.

Ochsner, Michelle, and Michael Greenberg. 1998. "Factors Which Support Effective Worker Participation in Health and Safety: A Survey of New Jersey Industrial Hygienists and Safety Engineers." *Journal of Public Health Policy* 19, no. 3: 350–65.

"Pain of the Kader Fire Won't Go Away." 2003. Trades Union Congress. *Risks*, no. 105, May 10.

"Professor Harry Glasbeek—Corporate Crime Fighter." 2003. OHS Reps Web site, VTHC, January 31. Available at: www.ohsrep.org.au/features/1043983091_10554.html.

Reiman, Jeffrey. 2001. *The Rich Get Richer and the Poor Get Prison*. 6th ed. Boston: Allyn and Bacon.

"Rep. DeLauro Presses OSHA for Tougher Criminal Enforcement." 2003. *Occupational Hazards*, May 2. Available at: www.occupationalhazards.com/full_story.php?WID=7019

"The 10th Anniversary of the Kader Fire in Thailand." 2003. *Occupational Health and Safety Rights*. Asian Network for the Rights of Occupational Accident Victims (ANROAV) newsletter, May. www.2bangkok.com/2bangkok/anroav/0305.shtml.

Trades Union Congress (TUC). 2003. "The Answers to Work-Related Deaths—Why We Need a New Corporate Killing Law." Corporate killing briefing, published by the Trades Union Congress, the Centre for Corporate Accountability and Disaster Action, March.

"Two Million Killed Each Year." 2003. *Hazards*, no. 81 (January–March): 22–23. Available at: www.hazards.org/haz81/ilodecentwork.htm.

"Unaccountable Accounting." 2002. *Focus on the Corporation,* January 25. Available at: www.lists.essential.org/pipermail/corp-focus/2002/000103.html.
"US Top Court Rejects Appeal over Cyanide Poisoning." 2002. Reuters, October 9. Available at: www.planetark.org/dailynewsstory.cfm/newsid/18109/story.htm.
"USA: Three Strikes and You're Out." 2003. Trades Union Congress. *Risks,* no. 105, May 10. See also Bill SB335, California State Legislature. www.info.sen.ca.gov/cgi- bin/postquery?bill_number=sb_335&sess=CUR&house=B&site=sen.
"Who Will Take Responsibility for Corporate Killing?" 2003. Editorial. *Lancet* 361, no. 9373 (June 7). www.thelancet.com/journal/v01361/iss9373/full/llan.361.9373.editorial_and _review.25985.1.
"The Writing's on the Wall." 2003. *Hazards,* no. 81 (January–March): 5.

3

Regulating Risk at Work

Is Expert Paternalism the Answer to Worker Irrationality?

Peter Dorman

The latest fashion in economics is the rediscovery of psychology, expressed in the nameplate "behavioral economics." If in the past economists routinely assumed perfect rationality and unlimited cognitive abilities among all workers, employers, and consumers, many are now willing to consider the consequences of an approach that looks more honestly at how people really think and act. This is widely viewed as a revolution in economic methodology, one that is just beginning to filter down into studies of public policy. Given the enormous influence of academic economics on respectable opinion in such fields as public health and labor relations, it is natural to ask what the impact of the new behavioral perspective will be.

But the behavioral economics of occupational safety and health is already in its second generation. This is because I relied to a great extent on early research in economic psychology in *Markets and Mortality: Economics, Dangerous Work, and the Value of Human Life* (Dorman 1996a)—a book that is in some respects a forerunner of current research, since it examined the consequences of replacing classical expected utility theory with cognitively based models of decision making.[1] Nevertheless, the approach I took is strikingly different from more recent studies, both in its selection of behavioral models and the interpretive framework it applies. Moreover, the book focuses primarily on the traditional economist's case for laissez-faire, and its behavioral analysis does not take cognizance of more recent arguments, inspired by behavioral economics, for paternalistic intervention.

Markets and Mortality had many other items on its plate: a critique of hedonic pricing methods used to derive values for life, an exposé of the excessive and unfairly distributed occupational risk that continues to plague the United States, and an analysis of policy alternatives.[2] These received the most attention. For whatever reason, the behavioral account of conflicts in the workplace was less interesting to most readers. Given the resurgence of work at the intersection of economics and psychology, however, I believe that account deserves a second look. My discussion in this chapter will not assume any familiarity with *Markets and Mortality,* but I will use its existence in the background as an excuse to be brief; the reader interested in a more complete story can consult that work.

I will begin with a short overview of the conceptual waves that have washed over the literature on occupational safety and health, emphasizing the interplay between explicit or implicit decision theory on one hand and policy advocacy on the other. Next I will take a closer look at the decision theory perspective of *Markets and Mortality,* particularly with respect to prospect theory and cognitive dissonance. In the third section, I will briefly outline the game-theoretic argument of *Markets and Mortality,* which attempts to marry the psychology of risk perception to the sociology of organizational behavior. I want to stress that the repeated-game approach to conflict over risk is not intended to justify all deviations from rational risk evaluation, and I hope I can make this point more effectively here. The fourth section shifts the ground to the implicit counterfactual in the current revival of paternalism, the presumed rationality of institutional ("expert") risk analysis. I will offer two reasons to doubt that this process offers the gold standard in decisional efficiency and lack of bias. I will then conclude with some brief thoughts on ways in which the law can be deployed to promote greater collective rationality, even in the face of individual-level bias and error.

"Law, Logic, and the Swiss Can Be Hired to Fight for Anyone," or Mustering the Troops

In a sense, the debate over hazardous working conditions has always centered on competing claims to logic. In early nineteenth-century England, the two poles of (educated) opinion on this issue were represented by Edwin Chadwick and the followers of Adam Smith.[3] For Chadwick, the problem was preeminently one of public health. Having conducted parliamentary inquiries on the matter, it was clear to him that the human toll in Britain's factories and workshops could hardly be efficient. The utilitarian calculus demanded a suitably engineered intervention that would eradicate those risks whose costs to society exceeded their benefits. The device Chadwick hit upon was the internalization of the costs of occupationally induced injuries and diseases—the application of the "producer pays" principle. For him, refusal to adopt such measures in the face of the evidence he had collected was stubborn irrationality.

But England did not follow Chadwick's advice. Far more influential were the jurists and political economists who had rallied to Adam Smith's side of the debate. For Smith, it was folly to second-guess the terms of any freely chosen agreement between competent adults, and this must include working conditions, whether dangerous or benign. Smith assumed that, in the course of their affairs, most individuals would rationally pursue their own interests, and thus any contractual outcome must be assumed to be acceptable to all parties. If some people acted unwisely, they would be instructed by the consequences of their actions to do better next time. Under the supposition that rational self-interest was the norm, Smith hypothesized that work that was more disagreeable than average would also receive above average compensation—why else would rational workers accept such jobs? Thus, workers in dangerous jobs should not be pitied or rescued; considering their monetary rewards, they were as well off as anyone else of their general social standing. This view, founded on the premise of rational choice by workers and employers, was highly influential in both British and American jurisprudence; it is the basis of the "assumption of risk" defense that saved many an employer from liability in the wake of catastrophic industrial accidents.

At the zenith of freedom of contract, Karl Marx was at work in the British Museum, devising a different sort of rationality to use as a weapon against the mill owners. In his version, the entire edifice of capitalism rests on the competitive drive to maximize profits, portrayed as the extraction of a surplus from the output of workers, the true producers. Since Marx accepted the Smithian critique of mercantilism, he believed that this surplus could be obtained not in the trade of goods, but only in the process of production itself. Thus the core of Marx's theory of profit had to do with working conditions: the length and intensity of the working day, the organizational and technological assault on workers' exercise of skill and autonomy, and the relentless pressure to reduce standards of safety and health. Perhaps few pages of *Capital* are as memorable as those that document the various parliamentary inquiries into hazardous and child labor; Marx takes these as evidence of an ineluctable race to the bottom characteristic of the capitalist mode of production.[4] Nominally, Marx is contemptuous of appeals to morality in the face of industrial carnage; they are, he argues, simply the "rationality" of a particular set of productive relations. To change the outcomes it would be necessary to change these relations. It is interesting to note, however, that Marx (who devoted hundreds of less-than-memorable pages in *Theories of Surplus Value* to Smith, Ricardo, and their lesser contemporaries) never challenged Smith's theory of equalizing wage differentials directly—this despite the important role the theory played in courts and Parliament during Marx's own lifetime. Perhaps he was so convinced by the logical structure of his argument that Marx did not see any room for effective worker choice.[5]

It was not Marx who turned back the tide of Smithian laissez-faire in the workplace, however. The rise of pragmatic social science at the turn of the twentieth century put a different spin on risk and rationality. This new approach, eventually incorporated into U.S. legal thought in the form of Realism, was predicated on a sociological framework in which individual choice was meaningful only in the context of social norms and institutions. Atiyah (1979), for instance, cites a dissent by Mellish, who asks why individual workers must choose working conditions on the basis of no presumption of employer duty, when it would be just as reasonable, and more socially beneficial, for worker choice to be exercised in a context of employer obligation. (This is roughly analogous to the modern distinction between choice over willingness to pay and willingness to accept.) Teddy Roosevelt, in a magazine article cited by Weinstein (1968), points to the ironic case of Sarah Knisely, who asked to have a guard placed on a machine she was tending. When her supervisor said no and threatened to fire her, Sarah went back to work and was subsequently injured. Her action against her employer was denied because her very complaint had demonstrated, according to the Supreme Court, that she was cognizant of the risks and had therefore accepted them. Such cases, seen through a Progressive lens, proved that the legal and institutional context of work had to be altered so that workers could exercise their choices to greater advantage. In other words, what was lacking was not Knisely's rationality, but a legal framework in which her rational actions could make her work safer rather than more dangerous.[6]

The pragmatic/realist view was dominant in occupational safety and health policy (at least in an aspirational sense) for more than a half century. It was eventually challenged by a resurrected Smithian orthodoxy in the 1970s, spearheaded, above all, by the work of W. Kip Viscusi.[7] At a high level of abstraction, there is little difference between the theoretical positions of Smith and Viscusi. Both rely on the rational self-interest of workers and employers to resolve differences over working conditions in the most efficient possible way, and both anticipate that such agreements will entail wage premiums for jobs imposing above average risks to life or health. Of course, mathematization, even when it serves to formalize prior nonmathematical arguments, has consequences. For one thing, it forces its practitioners to be explicit about matters that narrative theorists could conveniently evade. For our purposes, a central aspect of the algebraic formulation of wage compensation theory is that it required a firm grounding in expected utility theory. This was necessary if worker behavior was to be truly utility-maximizing in the face of a nondeterministic relationship between choice and outcome. It was also necessary for empirical testing, since the objective probability of fatal or nonfatal injury (insofar as these could be measured) was to be used as an explanatory variable, with behavior (reflected in wage rates) as the outcome to be explained. For the most part, the appropriateness of the classical model of expected utility was simply assumed by Viscusi et al., although some effort was made to justify it.[8] The practical implications of this approach were considerable. It resurrected the Smithian bias toward laissez-faire in risk regulation on both efficiency and equity grounds, but it also provided the foundation for new techniques that attach monetary values to life and health, thereby promoting the use of cost-benefit analysis (CBA) in policy realms having public health ramifications. This is still the orthodox position among economists at the time of this writing, and through the political ascendancy of economic analysis over competing policy and disciplinary traditions ("neoliberalism"), it has acquired wide social and political influence.

It is against this orthodoxy that behavioral studies of risk perception dissented. Although relevant psychological research can be dated from as early as the 1950s (inspired by the Allais and Ellsberg paradoxes in decision theory), the first behavioral analysis of occupational risk was Akerlof and Dickens (1982). The theme was again taken up by Dickens (1990) in his more expansive critique of hedonic labor market research. By the time I wrote *Markets and Mortality*, many of these ideas were in the air, due to the industry of behavioral risk analysis spawned by the rapid expansion of environmental regulation. One might think that the main policy inspiration for this revolt would be a return to the "higher" rationality of public risk management, but this was not generally the case. In the realm of risk policy, the central political debate during the 1980s and 1990s was whether CBA should play a determining role, and to a large extent this hinged on the ability of CBA to successfully incorporate such qualitative aspects of regulations as their effects on life and health. By undermining the theoretical underpinning for hedonic pricing techniques (and to a lesser extent the contingent valuation alternative), advocates of the behavioral approach were throwing sand in the gears of "rational" risk management. This had the effect of giving comfort to more politically responsive policy, since that was (and is) the perceived alternative to CBA.[9] In other words, the logical sequence went from a critique of orthodox expected utility theory, to a critique of the techniques used to monetize nonmarket values, to a critique of CBA in regulatory policy, to an endorsement of (relatively) unrestrained political determination of policy outcomes.

The behavioral law and economics approach to risk that has arisen since the publication of *Markets and Mortality* is, despite anatomical similarities, a beast of a different color. Its target is not primarily the rationality of economically inspired risk management, but that of the political process. In that sense, its roots are in the behavioral critique of the regulatory state found in the work of Douglas and Wildavsky (1982), as well as the attack on "environmental paranoia" undertaken by, among others, Ames and Gold (1999).[10] This transformation should not surprise us: if CBA is suspect because it derives a few of its valuations from the irrational choices of citizens, why should not the political process—which depends to a much greater extent on public rationality—be an even bigger target? Note that this observation applies equally to the workplace, where risk management is necessarily responsive, in some degree, to both public regulation and the risk perceptions of the workforce (the public at work). Thus, this new wave of behavioral analysis is skeptical of both laissez-faire and political intervention; it seeks a third way drawing on the rational expertise of public health professionals and economically literate policy analysts.

I hope this brief overview of the changing fashions in occupational safety and health policy has made it clear that rationality has always been a contested attribution, whether it has been thought to inhere in workers, the public health profession, the political process, or the economic mechanism generally. The content of rationality has evolved over time as well, and will no doubt continue to do so. It should also be noted that the conflict between the two most recent incarnations of the rationality principle—the rationality of democratic regulation vis-à-vis laissez-faire, and the rationality of technocratic regulation vis-à-vis the predictable irrationality of popular risk perception—do not exhaust the possible stances one may take on this set of issues. This is important: sometimes the impression is given that the most recent battle encapsulates the entire war.

The Behavioral Dimension of Risk

Imagine that a community of workers faces a risk factor that varies continuously over some range, from very low to very high. At the bottom of this range the risk is negligible; at the top it is dangerous in a way that would be recognizable by everyone in the community. Now, in addition, suppose that people have response functions that translate perceived risk into behavior, such as quitting a job or making health and safety demands in a unionized or similar workplace. If each individual's response function were unique, we would expect a community response function, representing the collective behavior of the entire exposed population, to vary as continuously as the risk level does: when risk is higher, the response is stronger and vice versa.

A different possibility is that the community displays *risk norms.* Such norms embody two types of discontinuity. First, individual responses are discontinuous around a particular level of risk. Under the assumption of continuity, if the risk increases slightly then so does the response. If there is a risk norm at work, however, a small change from risk below the normative level to risk above it can trigger a very large change in response. In other words, people act as if they have two categories, below-normative and above-normative risk, where changes within either category are less significant than the switch between them. Second, rather than having a range of risk norms (triggering levels), the community congregates on a single, shared level. Taken together, these two assumptions lead to the expectation that the community will turn on a dime, so to speak, with this dime being the risk norm. One common version of this situation arises when the norm is the lowest perceptible level of risk. In this case, the community responds as if it believes there should be no risk of this particular sort, reacting strenuously at the first sight of such a risk. While many risk norms take that form, they can also be found at other levels of risk, including those that most health professionals would deem to be too low or too high.

I took it as a stylized fact that such risk norms occur in the workplace, and in society as well. The source for this supposition is not the experimental literature, but the much richer, albeit less rigorously formulated, evidence drawn from the historical and institutional experience. *Markets and Mortality* contains a chapter that surveys such topics as the legal record, accounts of occupational safety and health (OSH) conflicts, and the supply and demand for regulation. Thus, the starting point for behavioral analysis was the need to explain why such norms exist and how they might be altered to promote more effective public health policies. I will describe the repeated game theoretic aspect of the explanation in the next section; here my focus will be on the aspects of the psychological literature that shed light on this question.[11]

Like more recent behavioral economists, I was drawn to the cognitive offsprings of the bounded rationality hypothesis. Mental accounting and selective attention heuristics were particularly important, but it is only in a social context, where contagion (or, as we will see, coordination) effects are relevant, that their implications for risk norms become apparent. Prospect theory also played a role, but it was given a different interpretation than the one we usually see today. In the contemporary literature, the discontinuous response to risk postulated by prospect theory has taken the form of status quo bias. The model proposes that individuals respond to potential outcomes by comparing them to a reference outcome, and that potential losses relative to the reference state are more salient than potential gains. This converges on status quo bias when the reference state is assumed to be the individual's initial position. There is a large experimental literature demonstrating that this is in fact the case, at least in situations constructed to test status quo bias. Indeed, in the stripped-down social contexts that characterize such experiments, it is hard to see how any other state could be salient enough to serve as a reference function.

In the real world, however, people move through social and cognitive environments in which attention is drawn to other, nonexistent states: "clean" (zero pollution) counterfactuals, maximum "acceptable" risks as designated by government regulations (such as threshold limit values), or risk levels that have been identified and publicized by advocacy groups. In such richer contexts, it seems appropriate to me to return to the original, general formulation of prospect theory that appears in Kahneman and Tversky (1979), where the reference state may be the status quo but is not restricted to it. Allowing for variable reference states has important implications for the sort of predictions we would make about worker (or public) behavior. They would include:

- The potential for nonconservatism. Status quo bias is inherently defensive in nature. Thus, in a recent paper, Sunstein (2002) analyzes the default rules incorporated in contract law in terms of their effect on workers' initial positions, which status quo bias would then induce them to defend. No doubt such effects do exist. There are many situations, however, in which workers and other parties to economic disputes set reference points well outside the domain of the status quo. In the conflict which rages over the use of genetically modified organisms (GMO) in the food supply at the time of this writing, for instance, many vocal members of the public are weighing food industry proposals against the standard of no GMO despite that the status quo is one of significant GMO presence. An ambitious reference state can be the lever that galvanizes public action and leads to significant changes in risk policy.
- The susceptibility to external influence. If reference states are tethered to the status quo, they can be changed only by changing the status quo. In a more general interpretation of prospect theory, however, reference states can be altered through persuasion, contagion effects, or other social mechanisms. This widens the range of potential strategic activity. In particular, it provides a useful framework for thinking about the role of advocacy organizing. Organizers, such as workplace or community activists, have as one of their main tasks that of bringing about a shift in

reference states, so that most individuals in the target population will regard the status quo as unacceptable. The point—the contribution of prospect theory—is that this does not require individuals to accept new information or to reevaluate actual or prospective outcomes, but only to adopt a new standard for comparison that situates the discontinuity in behavioral response. Such an analysis helps us to understand the volatility of worker response to occupational risk and the impact that union mobilization can play in arousing discontent with previously tolerated working conditions.

- The potential for interaction with other behavioral elements. Once the reference state is permitted to deviate from the status quo, additional explanation is required to determine what it will be. This opens space for constructing more complex behavioral models that draw on a variety of psychological mechanisms. An important example is the role that perceptions of employer intent sometimes play in workers' acceptance or rejection of hazards. It often happens that the level of risk provides, or is thought to provide, an indication of the employer's disposition; that is, whether the risk is largely inherent in the nature of the production process, or whether it is discretionary—excessive beyond what is feasible and profitable. A particular intensity of risk may be seen as occupying the dividing line between these two possibilities; it can therefore serve as a reference state. (I will return to the issue of perceived intent in the next section.)

While cognitive mechanisms related to bounded rationality have gained acceptance among recent students of risk behavior, it is surprising that the problem of cognitive dissonance has been largely ignored. This was the first brick thrown at expected utility theory in the economic literature on occupational risk (Akerlof and Dickens 1982), and it is essentially little more than a formal account of "denial," a universally recognized factor in the response to risk. For our purposes, the problem of cognitive dissonance arises in OSH contexts because awareness of risk exposure is unpleasant—in economic terms, a source of disutility. One way to explain its relevance is that attention to risk exposure, whether in the form of appropriation of information substantiating risk or mental effort expended toward contemplating this risk, is not only a means to an end (such as formulating a behavioral response that may lead to future increased utility) but also an experience in its own right, capable of conferring disutility. When individuals shield themselves from such information or conscious reflection in order to avoid this disutility, we say that they are "in denial." What is at stake in such situations is nothing less than the conflict between two different rationalities, one purely instrumental (the use of information and reasoning to make the most utility enhancing choices), the other a metarationality that incorporates the utility of cognition into the choice of behavior. Inasmuch as the second contains the first, it actually represents, in some sense, a more rational rationality. It is often, however, profoundly self-destructive.[12]

In practice, denial is ubiquitous where individuals and communities face significant risk. It is perhaps the foremost barrier to the dissemination of safer practices at work, frustrating worker activists and safety managers alike. Understanding the sources of denial should be a principal objective of social science analysis in the field of occupational risk; oddly, it has been neglected.[13]

In *Markets and Mortality,* I presented a simple model in which the net effect of cognitive dissonance is a function of the ability of agents to alter the circumstances that subject them to risk. Suppose that individuals can exercise a preconscious choice of whether or not to take account of a new piece of information that informs them of the risks they face. The cost of this perception is the cognitive dissonance it would cause: our hypothesized agents would experience mental discomfort in recognizing that they were exposed to factors that could cause them serious illness or early death. On the other side of the ledger is the benefit: the possibility that, by acting on this new information, they could reduce their level of exposure and the probability of physical anguish. As questionable as the utility maximization assumption appears in the context of conscious decision making, it seems even less supportable in the preconscious filtering process we are considering here—yet it may not be too far off the mark, and it makes for an interesting exercise. (Some sort of semirational preconscious filtering mechanism is required for cognitive dissonance to have any theoretical basis at all.) The cognitive discomfort can be assumed to be fixed, beyond the control of those afflicted by it or society in general. On the other side, however, the benefit of new information clearly depends on social arrangements. Take two extreme cases. In the first, those exposed to risk are completely powerless over the process of risk generation, either because the risk emanates from natural forces that humans cannot alter or because it has a social etiology, but those at risk are powerless within this social setting. Here there is no benefit to contemplating new information that heightens the sense of risk; it is all cost. In the other polar case, individuals have complete control over the degree of risk at no cost to themselves. Now it would be impossible for the costs of attending to new information to exceed its benefits, since the dreaded possibilities that trigger cognitive dissonance can be foreclosed through preventive action. (Eliminating cognitive dissonance would simply be one additional motive to exercise the power to expunge risk.) Presumably, most real world situations would be somewhere in the middle, with a less certain relationship between the costs and benefits of risk perception. Perhaps the most important factor in determining where on this spectrum a particular case lies is the relationship individuals at risk have to the power structure of the institutions that manage and allocate risk. In the workplace, this would most likely be the degree of employee involvement in decision making, either through participatory management structures or unionization. Thus, greater worker voice in the enterprise would tend to be associated with less influence of cognitive dissonance pressures—less denial of risk. Ironically, if the initial social position is one in which workers are poorly organized and have little clout, the effect of cognitive dissonance will be to discourage exactly the types of awareness that might, with a bit of collective organization, lead to greater voice and more honest perception of risk. How to break such a vicious cycle is the organizer's—and the public health advocate's—dilemma. Labor law can play a crucial role in such circumstances by giving workers ample personal control over hazardous conditions. In *Markets and Mortality,* I make the case for a greatly strengthened right to refuse imminently dangerous work, a right nominally incorporated in the Occupational Safety and Health Act. Surely its effect on workers' openness to new and potentially disconcerting information on occupational risk would be one of its chief advantages.

A third behavioral avenue explored in *Markets and Mortality* has to do with attitudes toward autonomy and attribution of risk. Although the literature in these areas was thin a decade ago, it seemed reasonable to point to two generalizations: individuals are more tolerant of risks they believe to be (partially) under their control, and they are less tolerant of risks that appear to be discretionary (self-serving) on the part of others who generate them. These are separate phenomena, but taken together they suggest a Kantian posture: depredations are valued not for their instrumental characteristics, but as transactions between subjects. In other words, I may refuse to bear a risk not because I believe it to have harmful consequences, but because I take it to be an instance of exploitation or maltreatment. Similarly, I may accept a risk even though it is harmful, either because I hold myself responsible for it, or because I regard it as a morally neutral act of nature. This would seem to be a centrally important issue in occupational safety and health, since risks on the job are subject to exactly this sort of mental evaluation. In a repeated game context (see the following section), this sort of behavior can be justified because it distinguishes between the behavior of others that has consequences for future interactions and that which does not. If resources I expend today can deter harmful actions others may take against me in the future, that increases their justification.

Since the publication of *Markets and Mortality,* however, the literature on the fairness dimension of risk and other economic goods or bads has expanded enormously. An important review by Fehr and Schmidt (2000) indicates that some earlier findings have been confirmed while others need to be revised. It is still the case, they report, that perceptions of intent matter greatly in experimental studies of economic behavior [illegible] taken in *Markets and Mortality*. Nevertheless, there are important qualifications. First, explanations of fairness effects based on models predicated on self-interest are undermined by evidence that concern for fairness survives experimental designs that eliminate any possibility of personal gain on the part of altruists. This result carries over to tests of individuals' willingness to punish unfair behavior, perhaps the more relevant issue in the context of workplace conflicts over risk. Second, individuals vary significantly in their orientation to fairness—a problem for attempts to use this aspect of behavior to explain the emergence of social norms surrounding risk and similar phenomena. Finally, for at least some subjects, fairness concerns appear to be deeply ingrained quite apart from any particular social setting that would entail repeated interaction. It appears, then, that in some respects *Markets and Mortality*'s approach to the analysis of risk in workplace and related settings is showing its age.

Too Much Static

There are many dubious propositions underlying conventional economic theory, but perhaps none is so dubious as the notion that transactions between individuals or between individuals and institutions are anonymous, nonrepeating affairs. This is at the heart of standard supply and demand analysis: the potential buyer approaches the potential seller; each has his or her offer curve in mind; and in an instant—in a period of time too brief to have any duration in theory—an exchange is consummated and the parties move on. Their calculative abilities do not extend beyond a direct appraisal of this one exchange for utility or profit. Of course, if anonymity is assumed, there can be no ramifications of today's choice for tomorrow's; econoland is a realm of perfect amnesia, where no one can remember who anyone else is or what they did.

It was in reaction to just this sort of issue that John Von Neumann and Oskar Morgenstern developed the apparatus of game theory. Market behavior as modeled by economists is just one of many types of behavior; it seemed to them that it was begging fundamental questions to assume that the tendering of offers, the communion of wills, and the instantaneous exchange of obligations is the substance of all economically relevant activity. The advantage of game theory is that it provides a syntax as flexible as the theorist allows, so that it can incorporate a far wider range of potential behaviors. For our purposes, it is important that game theory can model repeated interaction among agents:

myopic static optimization gives way to strategic, dynamic interaction. It is astonishing that anyone would try to model the employment relationship as other than nonanonymous, strategic, and repeated—yet the economics literature, including the literature on occupational safety and health, is replete with single-period static analysis. Such is the power of disciplinary convention.

It seemed to me that the most plausible representation of workplace disputes over safety and health conditions would be that of a repeated "chicken" game. In the simplest version of this game, there are two players (in this case, "workers" and "management"), and two potential moves for each side (impose/withdraw risk for management; accept/resist risk for workers). The "chicken" moniker derives from the death-defying game played by teenage drivers (and depicted famously in the film *Rebel Without a Cause*), in which the first driver to stop his car before the wall/cliff/so on loses. In such games, there is a temptation to unilaterally defect when the other player cooperates, but not to defect if the other player defects. If we translate cooperation and defection into our context of workplace safety, we get "there is a temptation for employers to impose greater risk if workers commit to nonresistance, but not if workers commit to resistance." (The symmetrical formulation does not make sense for workers, since there is no advantage to resisting risks that the employer has not imposed.) Thus, there are two potential equilibria in the stage game, workers commit to resistance and employers withhold risk or workers commit to acceptance and employers impose risk. In itself, this is rather uninteresting, but matters become more complex when we move to the repeated game context. Now the payoffs to each move must include the effects it may have on the other player's future strategies. It might well make sense, then, for either side to respond to the other's defection with its own, since the negative consequences of conflict in the short run may be offset by deterrent effects in the long run.

In a subsequent paper (Dorman 1996b), I demonstrated that, if payoffs to workers are continuous in the degree of risk selected by management, the move from a single-period game to a repeated game always leads to a lower trigger level for worker resistance. I also demonstrated that a Coasian solution, in which the risk transacted equilibrates the marginal cost to workers of being exposed to it and the marginal cost to employers of preventing it, would occur only by accident.[14] For the purpose of this chapter, however, the more relevant questions are, does a repeated game analysis contribute to the explanation of normative risk behavior, and does it support the case for paternalistic intervention in the regulation of risk?

It seems to me that the answer to the first question is yes, although in a loose rather than precise fashion. First, achieving the successful collective action required to be in the game—to be constituted as the collective player "workers"—entails coordination on risk response. In practice, this often leads to "bright line" solutions, highly accessible or clearly delineated risk levels that a community of individuals can utilize as benchmarks, so that they can reap the benefits from collective action. Examples of such norms might be government-promulgated regulatory standards, status quo exposures, de facto industry standards, or perhaps zero exposure (as in the case of persistent or bioaccumulative toxic agents). Second, perceptions of intent play an important role in repeated games, and particular exposure levels may stand at the boundary separating perceived discretionary and nondiscretionary imposition of risk. Once again, regulations may play this role, particularly if they are set according to a procedure that includes a feasibility test. Status quo exposure may also serve this function, since it provides evidence of the employer's viability. Whatever the specific reason, it is not difficult to incorporate assumptions in repeated games that give rise to discontinuities in the payoff/equilibrium outcome relationship.

As for the second question, much depends on how we interpret rationality and efficiency in the context of ongoing relationships, such as the employment relationship. To apply a static notion of efficiency appears to miss the point altogether, since this criterion is derived from static optimization methods. A truly dynamic method would yield not an equality between this marginal cost and that marginal benefit, but a more complex relationship that incorporates the reaction functions of players in response to each others' strategies.[15] To put the matter concretely, to know whether it is socially "efficient" for workers to reject or accept a particular level of risk, it would also be necessary to know how employers will process this information in their choices over future risks, workers' responses to those choices, and so forth. Thus, the anti-Coasian character of the game should not itself be a cause for concern, although outcomes are unlikely to adhere to any alternative conception of efficiency. External regulation has the potential for perverse effects so long as the regulatory body is unable to suppress workers' and employers' reaction functions.[16] Effective, if not perfectly efficient, policy would tend to take the form of interventions that tilt the balance of power toward the side whose interests are insufficiently respected, and that encourage such bargaining lubricants as transparency and trust. I will return to these points in the conclusion.

Consider the Alternative

In an imperfect world, the shortcomings of one institution or procedure can only be compared to the shortcomings of another. Unfortunately, the real-world fallibility of popular decision making in the context of risk is often compared to an idealized expert process, rational by assumption. The more interesting question is whether the irrationalities of worker and citizen behavior are more or less harmful than those of the institutionalized experts.

I will take it as given that experts are free of the behavioral tics we find in everyone else: their rationality is unbounded, and they are impervious to random environmental influences on their belief formation. Nevertheless, they still fall short of the gold standard in risk perception and response in at least two ways: they are swayed by self-interest in a world in which those who impose workplace risk have preponderant power, and the standard algorithm for rational decision making under uncertainty cannot be implemented.

First, consider the problem of self-interest. Despite outbursts of other-regarding behavior, such as those documented by Fehr and Schmidt (2000), it is normal to think that experts, like everyone else, are motivated by self-interest. While it is certainly the case that competence and at least the appearance of integrity are self-serving traits in most professional communities, it is also reasonable to expect that the concentrations of wealth and power typical of our society will have some effect as well. Professional careers can be advanced through access to monetary resources; less obviously, the problems that are viewed as important to solve are typically framed by economic and institutional elites. Finally, there is unlikely to be a Chinese wall between activities and beliefs: by working on particular sets of problems (where "success" is defined within the framework of the problem) and by pitching their work to elites, experts will find that, over time, their beliefs tend to adjust.[17]

There is ample evidence of this process in the field of occupational safety and health and in the related field of environmental toxicology. While it is not possible to subject the expert-bias hypothesis to a formal test, many of the examples are dramatic and of large practical importance:

- For decades, miners were systematically misinformed about the threats to pulmonary function due to the high concentration of particulates in the air they were breathing. The long struggle, in the face of industry control over medical research, to replace junk science with sound laboratory and epidemiological findings and to put regulation and compensation on the public policy agenda is documented in Smith (1987).
- The asbestos industry succeeded in suppressing research demonstrating the carcinogenic properties of this widely used substance, with the result that millions of workers were needlessly exposed to heightened risk of serious disease or early death (Selikoff and Lee 1978). Corporations have freely resorted to bankruptcy in order to limit their exposure to claims for compensation.

- Since the passage of the OSH Act in 1970, the threshold limit values (TLV) adopted to demarcate "unacceptable" risk from exposure to harmful substances have been delayed and weakened due to industry pressure. Many commonly used chemicals known to be hazardous to workers have no TLVs at all (Castleman and Ziem 1988).
- The dangers of occupational exposure to ionizing radiation have been systematically minimized by U.S. government agencies that generate this risk or are responsible for regulating it. Independent scientists who attempted to apply sound epidemiological methods to research in this field were criticized and risked losing their funding; only now, after decades of unnecessarily harmful exposures, is there an effort to set the record straight and compensate workers for the catastrophic health effects (Wald 2000).
- After lead additives in gasoline were introduced in the 1920s, the auto and petroleum industries successfully suppressed research findings confirming the hazards they posed to refinery workers and the general population. Industry-paid experts continued to defend the use of these additives long after the scientific basis for their prohibition was established. (Arguably, that basis existed before tetraethyl lead was first employed.) (Kitman 2000)

This list is hardly exhaustive; a detailed hazard-by-hazard survey would go on for several more pages.[18] It is important to remember that behind each of these episodes of excessive industry influence lies the complicity of highly trained physicians and research scientists, who willingly bent, twisted, and spun their findings to receive the benefits of corporate sponsorship, or to avoid the costs of being identified as a loose cannon or lone wolf. Of course, this history also records the courage of researchers who stood up to these pressures and defended the public interest—but in each instance there were long periods in which, despite individual acts of protest, science was successfully suborned.

Incidentally, there is nothing in this account to suggest that corporate interests are intrinsically more hostile to objective science than any other interests. Union-financed researchers may well have exaggerated the risks to health of potential stressors they investigated, and other interest groups could be expected to exert similar influence. What makes corporate pressures on science more consequential is that (1) they are biased toward underestimation of the risks to which workers are exposed, and (2) they are better financed and more politically ensconced, by a large margin, than those of any other segment of society.

Even when the research process proceeds free of external influence, the apparatus that conveys this research to the public—the general and specialized media, professional organizations, and the network of institutions that sponsor seminars, briefings, retreats, and so forth—can have the ability to misrepresent the state of the evidence. Thus, the science itself may be accurate, but the science as perceived by those responding to risk, such as workers, managers, and public policy makers, may be distorted. The example of Alar, while not primarily involving the workplace, is instructive.[19]

The public health "scare" in response to a 1989 CBS *Sixty Minutes* exposé of Alar, a chemical then sprayed on apples to achieve uniform ripening and lower harvesting costs, has entered the canon of risk perception anomalies. Hardly a litany of the public's behavioral quirks can be found that does not contain a mention of the paranoid stampede to shun apples, despite any evidence of significant risk. Here we can see the gullibility of the public and the pitfalls of populist risk regulation, or so the story goes.

The reality was quite different. Beginning in the 1980s, the Environmental Protection Agency (EPA), acting on the basis of several peer-reviewed studies, concluded that Alar was a probable human carcinogen. By the time of the television broadcast, two states had already banned the substance and apple growers in Washington State were promising to discontinue it voluntarily. Research subsequent to the "scare" confirmed the carcinogenic potential of Alar at effect levels well above those typically deemed actionable by EPA. This consensus was joined by federal agencies (EPA and the Public Health Service's National Toxicology Program) and international bodies (the World Health Organization's International Agency for Research on Cancer). The popular impression that the public was irrational to be concerned about Alar is not based on expert opinion but on a well-financed campaign of disinformation spearheaded by the American Council on Science and Health, an industry-funded public relations organization. This group successfully placed newspaper and broadcast "news reports," held seminars, and so forth. With sympathetic attention from opponents of environmental regulation in the media (including the American Academy for the Advancement of Science's organ *Science*) and academia, they were able to convey an impression about Alar exactly the opposite of the scientific consensus (Negin 1996; Environmental Working Group 1999). The moral of this story is that it is not enough to have competent, objective scientific appraisal of hazards in the workplace or the general environment, as difficult as this is to achieve; it is also necessary to have a disinterested mechanism for communicating these findings to decision makers. Succeeding on both counts is difficult.

Even if expert appraisal were unimpeachably disinterested, however, there would still be a bias in favor of more relaxed regulation of potential hazards. This second argument is particularly interesting in the context of this discussion, because it bears directly on the adequacy of the benchmark against which public behavior is measured—the model of rational choice under uncertainty. I will demonstrate that while this model is appropriate to static circumstances in which all information that will ultimately be known is presently available, it fails in the more realistic case in which new and unexpected information is acquired over time.

Consider the standard model in the context of a decision whether to allow worker exposure to a chemical at a particular level of concentration.[20] After a series of studies are performed, we can construct a probability function for harm at various concentrations. Thus, we might conclude that, at the concentration we are considering, there is .001 probability per worker-year of exposure of contracting a specified disease. Suppose that this level is determined to lie exactly at the most appropriate TLV.[21] In this way scientific research is used to set regulatory parameters.

But it turns out that our experiments were not exhaustive, as they seldom are. A few years later new studies are undertaken based on different methods and assumptions, and we learn that, using the same .001 probability standard, the TLV should be revised downward. This type of event, which is the norm in regulatory policy over time, might be invoked to demonstrate that the system "worked": it was responsive to evolving science and altered yesterday's standard, rational in light of the available data, to today's new, improved model. A sequence of such standards, each derived from its supporting scientific basis, is viewed as a measuring rod to which the imperfect decision methods of the general public are compared.

In retrospect, however, it is important to stress that the first standard was too low. We did not know this at the time, but it turned out to be the case. Such errors are, of course, unavoidable; that is what it means to have incomplete information. Nevertheless, not *all* such errors are unavoidable. Specifically, errors based on partial information are justified only if they are randomly distributed around the standards we choose to set. In other words, there should be an equal likelihood that the next round of studies will cause us to raise the TLV as to lower it. This is a basic principle of probability theory and is obvious on reflection: If a calculation is made on the basis of all available information, including current information that predicts new information, then new knowledge will be a true surprise, with an unpredictable effect on our future calculations.

Two things should be apparent from this argument. First, nearly all existing regulation violates this principle. The history of setting occupational safety and health standards is replete with examples of TLVs and other designations that have been tightened over time and few that have been loosened. Far from representing an efficient response to the accumulation of scientific evidence, these trends indicate that regulatory agencies err on the side of lenience, repeatedly and systematically. It is this bias, rather than an idealized depiction of risk assessment based on full information, that should be compared to popular perception.

Second, the correct response to knowledge that evolves over time will involve considerable exercise of "soft" judgment, one that looks more like guessing than adherence to formal models of rational decision making. In order to set standards that have the property of random response to (random) new information, regulators will have to pay attention to indicators that hint at the likelihood of future evidence pointing in one direction or the other. This would involve taking into consideration existing results with relatively large Type I (false positive) error, contrary to the imperatives of "sound" science, and even informed hunches based on the history of this or similar fields of research. While this more expansive process is easy to justify from the standpoint of information theory (all information, and not just that which passes science's test of Type I error minimization, has value in rational decision making), it will look quite different from the ordered, formally representable methods against which the "nonrational" judgments of the general public are compared. Indeed, it remains an open question whether decision heuristics that have been criticized as nonrational, such availability (the extent to which a risk has been vividly brought to attention), are in fact related to the likelihood that the risks they address will be seen as more consequential over time. In other words, people may use signals like mass media coverage as a rough indicator of the extent to which future research is likely to focus on a risk and provide more grounds for regulating it.

The more general point about expert versus public reasoning that this analysis points to is this: all of us are inclined to believe things that are not true on the basis of small sample bias, noisy data receptors, and the other flawed procedures that science is pledged to abjure. Nonetheless, many of the perceptions that derive from first-person experience are at least partially accurate, and they have the potential to capture truths about our world long before they can be corroborated by the formal methods of statistical or laboratory analysis. In this way, workers may well be ahead of formally trained experts in acquiring information that predicts future research results, as is required if regulatory bias over time is to be avoided. Here is another argument for the practical importance of tacit knowledge (Polanyi 1958).

There is a long history of such popular quasi knowledge playing an important role in the workplace, serving to offset the limitations of expert opinion. The miners resisting the black lung epidemic drew on their personal experience in the face of medical opinion denying the occupational etiology of the disease (or even the existence of it). More recently, workers suffering from low-level exposures to multiple chemical agents have met with rejection from most of the scientific community, but their physical reactions persist. It is probable that at some future date at least some of their concerns will be validated by replicable research (Ashford and Miller 1998). In the meantime, the pressure arising from workers acting from direct experience can be an important corrective to the downward bias of "rational" risk assessment.

Conclusion

Each approach to resolving conflicts over dangerous working conditions has raised the banner of "rationality." Risks should be left to the market, or to the workers, or to the aroused public, or to the experts based on one or another definition of rationality and identification of its owners. Even today, much of the analysis conducted under the influence of research on nonrational psychological mechanisms measures itself against an Olympic ideal of unbiased, optimally efficient rational choice. While there are specific aspects of rational methods that should be defended on their own merits, I am arguing that the rational/irrational framework is profoundly misleading.

I have advanced two reasons for this. First, the rationality applicable to decisions taken in isolation from one another is not the same as the *strategic* rationality that should enter into decisions linked to one another through their embeddedness in social institutions such as the workplace. Not all of the departures from classical rationality (the expected utility model) displayed by individuals in laboratories and in the field are due to strategic complexities, but many are. It is reasonable to focus on risks around which collective action can be organized and for which there is evidence of purposeful intent on the part of the risk creators. It is not reasonable, however, to yield to the transitory comfort of denial, when new information could lead to real reductions in risk exposure.

Second, the shortcomings of popular responses to risk must be measured against the evident biases of expert risk assessment. There are valid arguments for intervention in occupational safety and health, of course, to remedy imbalances of power and the frequent lack of information on the part of both workers and companies. All industrialized countries deploy regulatory apparatuses for such purposes. But the goal of regulation should not be to substitute the assessments of experts for those of the workforce or the general public, absent deficiencies of information or power. If expertise is not persuasive on its own merits, there may be strategic or epistemological considerations to account for it.

An analysis such as this suggests an important role for the law, greatly underutilized in the United States. If decisions surrounding risk exposure are made in the course of intraorganizational bargaining, there is much to be gained by crafting rules that make this bargaining relatively efficient and open to rational persuasion. Risk regulation, aside from establishing minimum acceptable exposures, should be concerned with workplace reforms that provide for greater information sharing (such as joint OSH committees and right-to-know guarantees) and trust building (joint governance over other issues, more general disclosure requirements). Transparency can also be fostered by utilizing the "action plan" methodology pioneered in Denmark, which provides for extensive stakeholder input into OSH decision making (Karageorgiou et al. 2000). Greater stability in employment relations, reversing the recent trend toward flexibilization and higher turnover, can also be cooperation enhancing. Finally, measures to strengthen worker voice, including amendments to the National Labor Relations Act that would facilitate unionization and possible innovations such as works councils and codetermination, would not only provide for greater balance in bargaining outcomes, but would reduce the incentive for workers to submit to the pressures of cognitive dissonance.

Individual and collective rationality are not the same, either in process or content. The highest purpose of the law is to promote, in John Dewey's sense, a collectively rational process of learning and deliberation in all institutions, including the workplace. While such a process may be constrained by the individual folly we bring to it, it also holds out the possibility that there may be a little less folly at the end of the day.

Notes

1. Dickens (1990) is the only predecessor I know of.

2. For updates on the critique of hedonic methods and policy analysis, see Dorman and Hagstrom (1998) and Dorman (2000), respectively.

3. The following account is drawn from Atiyah (1979). Chadwick has attracted more attention in recent years; see Brundage (1988) and Hamlin (1998). I offer my own speculation on why Chadwick's approach to occupational risk has gained renewed currency in Dorman (1997).

4. These pressures were modeled as the competitive depletion of a common property resource—a healthy working class—by Wolfson, Orzech, and Hanna (1986).

5. A notable attempt to flesh out this dominance of structure over agency in the Marxist theory of working conditions (and elsewhere) is Cohen (1995).

6. A cogent statement of the strategy of pragmatic social science is Commons (1931); it deserves to be reread. He argues for a combination of individual and collective action in the application of rationality to social problems.

7. The true engineer of Smith's revival may have been Sherwin Rosen, who developed the mathematical model used to represent wage compensation theory in modern dress, and who was also the mentor of Richard Thaler, the first economist to attempt an empirical estimation of the extent of wage compensation. Viscusi's mentor was Richard Zeckhauser of the Kennedy School—and not Ralph Nader, for whom he worked on a study of western water policy.

8. See Viscusi and O'Connor (1984). The willingness to accept expected utility theory in the absence of detailed investigation may be due to the Friedmanite methodological prescription: if a theory's inferences work empirically, do not worry about tests of its assumptions. Positive coefficients on risk in wage-risk regressions were viewed as exactly this sort of inferential confirmation.

9. For an eloquent statement of this position, see Sagoff (1988).

10. There is an interesting logic to the appeal of psychological argument to scientists skeptical of the environmental movement. Their main intent is to dispute environmental concerns on technical grounds, but to bolster their argument they need to identify the motive underlying "erroneous" thinking. If the case against most environmental regulation is as conclusive as they believe, why do citizens persist in demanding it? Self-interest is not plausible, at least not for most of the general public, so some other explanation must be found. The catalog of psychological error provided by psychologists of risk perception fills this role nicely.

11. A second motive for both the psychological and game-theoretic investigations was to explain why worker behavior does not conform to the predictions of compensating wage differential theory. I will not pursue that issue in this chapter.

12. One way to reconcile these two considerations is to regard the tendency to indulge in excessive denial as a species of the general problem of weakness of will, the inability to transcend the salience of immediate experience in order to give reign to less myopic preference. For a variety of models that seek to explain this phenomenon, see Elster (1986).

13. It can be argued that cognitive dissonance mechanisms are at the heart of two texts in the Western political theory canon that address the role of unfree individuals in perpetuating their unfreedom, Etienne de la Boetie's *Discourse on Voluntary Servitude* (New York: Columbia University Press, [1548] 1942), and Albert Memmi's *Dominated Man: Notes Toward a Portrait* (Boston: Beacon Press, 1968).

14. In this article, the game is played between environmentalists and a government agency proposing a polluting project, but the analysis is formally identical.

15. Formally, these would be the first-order conditions of a Hamiltonian incorporating all the strategic interactions embodied in the repeated game.

16. This point echoes Coase's critique of Pigovian taxes and subsidies in a zero transaction-cost environment—an environment in which the preferences of the parties to engage in bargaining can be realized.

17. This is a greatly oversimplified summary of the modern approach to the theory of ideology. For a richer account, see Barnes (1977).

18. See also the discussion in Krimsky (2003).

19. Of course, any substance sprayed on agricultural produce is a potential occupational hazard for field workers but this was not the focus of the debate over Alar.

20. This account summarizes the argument in Dorman (2004).

21. There are pitfalls in this process, but I will assume here that they do not impede the analysis.

References

Akerlof, George, and William T. Dickens. 1982. "The Economic Consequences of Cognitive Dissonance." *American Economic Review* 72, no. 3: 307–19.

Ames, Bruce, and Lois Swirsky Gold. 1999. "Pollution, Cancer and Food: Science's Sifter Applied to Nine Misconceptions." *La Recherché* 30, no. 324: 47–53.

Ashford, Nicholas, and Claudia S. Miller. 1998. *Chemical Exposures: Low Levels and High Stakes*. New York: Van Nostrand Reinhold.

Atiyah, P.S. 1979. *The Rise and Fall of Freedom of Contract*. Oxford: Oxford University Press.

Barnes, Barry. 1977. *Interests and the Growth of Knowledge*. London: Routledge and Kegan Paul.

Brundage, Anthony. 1988. *England's "Prussian Minister": Edwin Chadwick and the Politics of Government Growth, 1832–1854*. University Park: Pennsylvania State University Press.

Castleman, Barry I., and Grace E. Ziem. 1988. "Corporate Influence on Threshold Limit Values." *American Journal of Industrial Medicine* 13, no. 5: 531–59.

Cohen, G.A. 1995. *Self-ownership, Freedom, and Equality*. Cambridge, UK: Cambridge University Press.

Commons, John R. 1931. "Institutional Economics." *American Economic Review* 21, no. 4: 648–57.

Dickens, William T. 1990. "Assuming the Can Opener: Hedonic Wage Estimates and the Value of Life." *Journal of Forensic Economics* 3: 51–9.

Dorman, Peter. 1996a. *Markets and Mortality: Economics, Dangerous Work, and the Value of Human Life*. Cambridge, UK: Cambridge University Press.

———. 1996b. "Norms and Regulation in Environmental Disputes." Unpublished manuscript.

———. 1997. "Internalizing the Costs of Occupational Injuries and Illnesses: Challenge or Chimera?" Presented at the European Conference on the Costs and Benefits of Occupational Safety and Health, the Hague.

———. 2000. "The Economics of Health, Safety, and Well-Being at Work: An Overview." Geneva: International Labor Organization. Available at: www.ilo.org/public/english/protection/safework/papers/ecoanal/ecoview.htm

———. 2004. "Evolving Knowledge and the Precautionary Principle." Unpublished manuscript.

Dorman, Peter, and Paul Hagstrom. 1998. "Compensating Wage Differentials for Dangerous Work Reconsidered." *Industrial and Labor Relations Review* 52, no. 1: 116–35.

Douglas, Mary, and Aaron Wildavsky. 1982. *Risk and Culture*. Berkeley: University of California Press.

Elster, Jon, ed. 1986. *The Multiple Self*. Cambridge, UK: Cambridge University Press.

Environmental Working Group. 1999. "Ten Years Later, Myth of 'Alar Scare' Persists." www.ewg.org/pub/home/reports/alar/alar.html.

Fehr, Ernst, and Klaus M. Schmidt. 2000. "Theories of Fairness and Reciprocity—Evidence and Economic Applications." CESifo Working Paper Series No. 403.

Hamlin, Christopher. 1998. *Public Health and Social Justice in the Age of Chadwick: Britain 1800–1854*. Cambridge, UK: Cambridge University Press.

Kahneman, Daniel, and Amos Tversky. 1979. "Prospect Theory: An Analysis of Decisions Under Risk." *Econometrica* 47, no. 2 (March): 263–91.

Karageorgiou, Alex, Per Langaa Jensen, David Walters, and Ton Wilthagen. 2000. "Risk Assessment in Four Member States of the European Union." In *Systematic Occupational Health and Safety Management: Perspectives on an International Development,* ed. Kaj Frick, Per Langaa Jensen, Michael Quinlan and Ton Wilthagen. 251–84. Amsterdam: Pergamon.

Kitman, Jamie Lincoln. 2000. "The Secret History of Lead." *The Nation.* Available at: www.past.thenation.com/issue/000320/0320kitman.shtml

Krimsky, Sheldon. 2003. *Science in the Private Interest: Has the Lure of Profits Corrupted Biomedical Research?* Lanham, MD: Rowman & Littlefield.

Negin, Elliot. 1996. "The Alar 'Scare' Was for Real." *Columbia Journalism Review* September–October: 13–15.

Polanyi, Michael. 1958. *Personal Knowledge: Towards a Post-Critical Philosophy.* Chicago: University of Chicago Press.

Sagoff, Mark. 1988. *The Economy of the Earth: Philosophy, Law, and the Environment.* Cambridge, UK: Cambridge University Press.

Selikoff, Irving, and Douglas Lee. 1978. *Asbestos and Disease.* New York: Academic Press.

Smith, Barbara Ellen. 1987. *Digging Our Own Graves: Coal Miners and the Struggle over Black Lung Disease.* Philadelphia: Temple University Press.

Sunstein, Cass. 2002. "Switching the Default Rule." *New York University Law Review* 77, no. 1: 106–34.

Viscusi, W. Kip, and C. O'Connor. 1984. "Adaptive Responses to Chemical Labeling: Are Workers Bayesian Decision Makers?" *American Economic Review* 74, no. 4: 942–56.

Wald, Matthew L. 2000. "U.S. Acknowledges Radiation Caused Cancers in Workers." *New York Times,* January 29.

Weinstein, James. 1968. *The Corporate Ideal in the Liberal States: 1900–1918.* Boston: Beacon Press.

Wolfson, Murray, Z.B. Orzech, and Susan Hanna. 1986. "Karl Marx and the Depletion of Human Capital as an Open Access Resource." *History of Political Economy* 18, no. 3: 497–514.

II

Old and New Challenges to Occupational Safety and Health in the United States

4

Silicosis and the Ongoing Struggle to Protect Workers' Health

Gerald Markowitz and *David Rosner*

In 1991 when we published *Deadly Dust,* our history of silicosis, a dread occupational lung disease caused by the inhalation of finely ground sand, we thought we were developing an interesting case study of the social relationships that allowed for the identification and the "forgetting" of an obscure disease. It was a book, we thought, for a small group of specialists in the history of medicine and we hoped it would become a model for those investigating the social history of public health. But since that time, silicosis has emerged as a major occupational problem and has attracted the Occupational Safety and Health Administration (OSHA), the National Institute of Occupational Safety and Health (NIOSH), the Mine Safety and Health Administration (MSHA), along with other agencies to name it as their target disease. Our book has played a role in bringing attention to this disease again and has also played a role in lawsuits and government meetings. Recently, different newspapers and legal journals have noted that silicosis may be the "new asbestos" (Glater 2003).

We did not intend to write a history of silicosis when we began our project in 1983. Our goal was to write a history of occupational and environmental health because few historians were paying any attention to the enormous impact on workers' lives of diseases created by the industrial workplace. In fact, many activists argued that there was no history of occupational and environmental health prior to OSHA, the Environmental Protection Agency (EPA), and Rachel Carson. We thought this project would last approximately three years.

However, it quickly became apparent that this history was indeed complex, deep, and very interesting. As we researched the general topic it became clear that silicosis—a disease we and most other historians (and indeed, most physicians) had never heard of—had been a major issue. As we interviewed older physicians about black lung and asbestosis (which dominated the news at that time), they continually referred to silicosis in the 1930s as a critical moment in their lives and in medical and public health thinking about occupational lung disorders. This observation coincided with our growing pile of articles and reports on silicosis in the first third of the twentieth century that had accumulated in our office. The combination of the growing stack of materials in the corner and the comments by older occupational doctors made us realize that the story of occupational and environmental disease was impossible to tell without an understanding of this seemingly obscure condition.

The book we wrote tells the story of the social construction and deconstruction of a condition that dominated public health, medical, labor, and popular discourse on disease in the 1930s, but virtually vanished from popular and professional consciousness after World War II. How, we asked, could a chronic disease that took decades to develop and was assumed to affect hundreds of thousands of American workers in one decade disappear from the literature and public notice in less than a decade? This question is the basis for *Deadly Dust* and we believe we provided a cultural, medical, and political model of how we, as a society, decide to recognize or forget about illness. The book received widespread praise in historical and medical journals, being called a "paradigm" for historical research on disease. The story we told tapered off in the 1950s and early 1960s when the literature on silicosis virtually stopped. As historians, we believed this was the end of our story (Rosner and Markowitz 1991).

While we were wary of traditional explanations about why the disease vanished (i.e., that it was "cured" or was eliminated as a problem by preventive measures), we implicitly accepted the idea that this was, in fact, a disease of the past. We quickly learned otherwise when, shortly after the appearance of the book, lawyers in Texas, Louisiana, New Jersey, and elsewhere began to call us to appear as expert witnesses in silicosis cases. We soon learned of horrifying situations today that made us realize that far from being a "disease of the past" silicosis unfortunately was quite alive. Migrant Mexican workers—both legal and undocumented—were dying in the oil fields of West Texas where they had been brought to sandblast with no protection the insides of various oil tanks, pipes, and other equipment used when the oil shortage of the 1970s led to the rejuvenation of this industry. In Louisiana ship builders and painters were coming down with this disease; elsewhere, in a host of foundries and other industries, workers were regularly being diagnosed with silicosis.

Dramatic and very troubling stories began to accumulate as we got deeper into these cases. We learned of a physician in West Texas whose life was virtually destroyed when he began to diagnose cases in Mexican American workers. He was forced out of the medical society, denied his hospital privileges, and lost his friends and colleagues in Odessa and Midland, all because he appeared to be pinning blame for these workers' deaths on the oil companies that dominated the economic and social life of the area. The workers' stories in and of themselves were heartbreaking, as we visited the area and learned of families destroyed by the slow, painful, and inevitable death from this disease. John Farmer, an African American sandblaster, was born during the Depression along the Gulf Coast of Texas.[1] He spent two years in college, joined the army, and then worked primarily in a shipyard as a laborer and sandblaster until 1982, when he retired. Because of his short, thin stature (five feet, six inches tall, weighing less than 125 pounds), Farmer was often sent into the poorly ventilated holds and double bottoms of ships, where he sandblasted off asbestos and other residue. He usually used a "desert hood" to protect himself against ricocheting particles, a cartridge respirator to partially filter the silica-laden air he breathed, and sometimes an air-fed hood: a cumbersome, spacesuit that supplied him with relatively pure air for the short time he was able to wear it in the hot and humid environment of this Southern shipyard. In 1988, when he was fifty-three years old, a doctor diagnosed him with "massive progressive fibrosis." Three years later, he had deteriorated to the point that he "was no longer able to walk and [could] only stand briefly while using supplemental oxygen." After considering him for a lung transplant, the physician noted that this fifty-six-year-old man's future looked "bleak," in terms of both longevity and quality of life (Personal communication, M. Diane Dwight, Provost & Umphrey, Beaumont, Texas, 1997).

Lawrence Brown was born in Louisiana in 1946. Following discharge from the army in 1977, he began working as a sandblaster and painter for a company that contracted nonunion workers out to almost every major refinery in the Port Arthur, Texas, area. Brown usually wore a desert hood and sometimes a paper dust mask as he blasted the insides of storage tanks and other vessels, preparing them for painting. In 1988, at the age of forty-one, he developed night sweats, violent coughing, and shortness of breath. He was diagnosed with tuberculosis at a veterans' hospital. Two years later he had lost more than twenty pounds and had persistent coughing, intermittent episodes of vomiting, and shortness of breath. He had difficulty exerting himself to take a shower. In June 1990, doctors reevaluated his X-rays and diagnosed silicosis. Brown died at the age of forty-six. He had sandblasted for only ten years (Personal communication, M. Diane Dwight, Provost & Umphrey, Beaumont, Texas, 1997).

From the 1990s to the present, thousands of lawsuits have been filed across the country by lawyers for workers in a host of "dusty" industries that have reawakened national attention to the ongoing threat from silica exposure. Through the discovery process of the lawsuits, thousands upon thousands of new documents have been uncovered about the machinations of industry and government in the years following the formation of OSHA, NIOSH, the EPA, and MSHA in the early 1970s. The industry mobilized to stop any reform from happening. We have in our possession thousands of pages of material from the Silica Safety Association (SSA) and other industry groups dedicated to stopping efforts to tighten silica exposure safety standards and to once again keep silicosis out of the public's purview.

Because we have been called to testify in depositions and in court about the longstanding knowledge of the industry concerning the dangers of silica exposure to the workforce, we have continued to follow the story and to gather new information about these efforts by industry. We are particularly proud that lawyers for various industries have sought to get judges to exclude *Deadly Dust* in court cases. Most recently, a judge in Mississippi was asked by defense lawyers to specifically ban any use or mention of our book in court. The request was denied.

The Debate Over Banning Sand

The story of silica's reemergence as a public issue begins in the late 1960s when sandblasters, painters, and other workers at the Avondale Shipyards in Louisiana began to come forward, complaining of constricted breathing and terrible pain. Scientists Morton Ziskind, Hans Weill, and Bezhad Samimi at Tulane University in New Orleans began a series of epidemiological studies that documented widespread silicosis among workers at nearby Gulf Coast shipyards. This documentation of the silicosis epidemic in Louisiana coincided with the passage of the Occupational Safety and Health Act of 1970, which created the NIOSH in the Department of Health, Education and Welfare and the OSHA in the Department of Labor. NIOSH was established to develop scientifically sound standards for occupational hazards while OSHA's mandate was to enforce safe and healthy work practices.

One of NIOSH's first activities was to produce reports that would provide scientific justification for OSHA's regulatory activity. Silica was among the first substances that NIOSH examined. Because silica was one of the oldest and presumably best documented occupational diseases, NIOSH believed, somewhat naively, that developing the scientific base for an enforceable standard would be politically less controversial than establishing standards for newer toxic substances (Personal communication, John Finklea, MD, NIOSH, June 1993).

In the early 1970s, partially in response to the Tulane studies, NIOSH sponsored an independent investigation of sandblasting practices throughout the country. NIOSH contracted with Austin Blair, an industrial hygienist from the Boeing Aerospace Company in Seattle, whose report proved to be an indictment of what industry then deemed to be safe silica exposures and of the lack of protection that respirators and protective equipment afforded workers. In the 1950s and 1960s it had been assumed that silicosis could be prevented if workers used respirators that lowered inhaled dust to levels below the threshold limit value (TLV) specified by the American Conference of Governmental Industrial Hygienists (ACGIH) and adopted by OSHA. But Blair's report raised questions about this assumption; he found that with the equipment then commonly used, "the protection afforded these workmen is, on the average, marginal to poor" (Blair 1974).

The indictment of protective equipment was serious enough. But, shortly thereafter, in mid-1974, two other NIOSH-supported projects reported on cases of silicosis among shipyard workers and steel fabricators in New Orleans and elsewhere (Ziskind 1974; Goodier et al. 1974). The response of NIOSH officials was swift; in 1974 the agency issued a Criteria Document Recommendation for a Crystalline Silica Standard, which recommended that OSHA's legally enforceable standard be made more stringent by cutting it in half to fifty micrograms per cubic meter of air. The document further recommended that silica be banned as an abrasive in blasting (NIOSH 1974).

In February 1975, just after the NIOSH document was published in the *Federal Register,* more than fifty people representing the affected industries gathered together in Houston to form the SSA. Their stated goal was to "investigate and report on possible health hazards involved in [the] use of silica products and to recommend adequate protective measures considered economically feasible" ("Report of Preliminary Meeting" 1975), but in reality their purpose was to make sure that OSHA did not adopt the NIOSH recommendation. Shortly thereafter, SSA wrote to various industries requesting financial support for the organization and that letters be written to the OSHA Docket Officer requesting delay in the public hearings on NIOSH's proposed standard. In these appeals it was clear that the primary purpose of the organization was "to represent interested parties in the attempt to assure the continued use of sand in abrasive blasting operations" ("Sline to Dear Sir" 1975).

In the course of the next few months, SSA developed its argument justifying the continued use of sand. L.L. Sline, the organization's president, argued that sandblasting was safe ("More on Proposed" 1975). SSA's position was that if workers used "proper protective devices" there was little danger of excessive exposure. Therefore, if equipment was capable of lowering exposure to "safe" levels, there was no need to lower the existing TLV for silica. SSA further argued that the reason that workers in the past had come down with silicosis was that they "had no air-fed hoods" (Wright 1977). What SSA officials did not reveal, however, were the results of a study conducted in a plant owned by one of the officers of the association, which showed that "under conditions considered good work practice," nearly half of all air samples were above the TLV, indicating danger for workers (Silica Safety Association 1977).

SSA was successful in delaying OSHA's adoption of the NIOSH recommendation (Wright 1981). In 1977 President Jimmy Carter appointed Eula Bingham to head OSHA. During her tenure more occupational safety and health standards were promulgated and adopted than in any similar period before or since, but no new silica regulations were adopted and none have been to the present day (Markowitz and Rosner 1995).

The final blow to the NIOSH proposal to ban sand came when Ronald Reagan was elected president. As the executive director of SSA observed: "With the change in administration, the ever increasing avalanche of government regulations have been reversed. Economic impact studies are now a required part of every regulatory process. As a result, OSHA's proposed abrasive blasting standard has been moved from a top priority 'target' regulation to the back burner" (Wright 1981, 184).

By 1982 the antiregulatory and probusiness environment in Washington had all but killed the efforts to lower the silica standard and made lobbying efforts unnecessary. With its success, SSA found its contributions drying up (SSA Newsletter 1982). A few months later, a special meeting of the SSA's Board of Directors concluded that "the association should be put on hold" (Wright 1983). The records of the organization were placed in storage and the offices closed.

A New Epidemic Emerges

Ironically, at the very time SSA was winding down its operations in East Texas, the silicosis epidemic was spreading from the Gulf Coast region to the West Texas oil fields in the Permian Basin, around Odessa and Midland. The West Texas oil fields, which had been in a long period of decline, began to boom as domestic oil prices rose as a result of the oil crisis in the mid-1970s. In the process of restarting the fields, workers reconditioned and cleaned miles and miles of piping, scores of small and large oil storage tanks, and a large amount of equipment.

The oil companies generally hired small, nonunion contracting companies to blast off tar and oil residues that had accumulated over the years in pipes and storage tanks used to store the raw oil product. Most of those employed to do the dirty and extremely dangerous job of sandblasting were Mexican migrants who had recently arrived in the boom towns of West Texas. Many of the workers were never provided more than bandannas or desert hoods to place around their noses and mouths to shield them from the silica sand they blasted. Desert hoods protect the heads and chests of workers from ricocheting particles but do nothing to protect workers' lungs from dust-laden air. In the early and mid-1980s, young men began to appear in doctors' offices complaining of shortness of breath, coughing, and sweating—symptoms of silicosis. Anti-immigrant feeling masked this outbreak until eventually the prolonged oil boom in the region brought silicosis to the attention of occupational health physicians and federal and state occupational safety and health agencies.

In November 1988 the epidemic that had been brewing for the past decade became public when the Ector County, Texas Health Department was informed by a physician in Odessa, a nearby oil town, that he had diagnosed three men as suffering from acute silicosis. Within a year seven other sandblasters had been identified as victims of silicosis. All were Mexican Americans, seven of them under the age of thirty, and all had worked cleaning tanks and pipes used in the oil fields. By the early 1990s, scores of workers, mostly Hispanic and mostly young, had come down with the disease.

In the late 1980s and early 1990s workers exposed to silica began to sue sand providers and equipment manufacturers that were not protected by workers' compensation statutes developed in the 1930s to limit liability suits against employers. Under Texas common law, those selling dangerous products had an obligation to adequately warn users of potential hazards. Using legal strategies developed in the asbestos litigation of the mid-1970s and 1980s, plaintiffs' lawyers began to reach substantial settlements, and many of the companies began to substitute nonsilica abrasives for the deadly sand previously used.

The election of President Bill Clinton in 1992 led to an attempt to revitalize NIOSH and OSHA. Within OSHA, people such as Mike Connors and Richard Fairfax who had long been concerned about silicosis were given positions of authority. Within NIOSH, Gregory Wagner was director of the Division of Respiratory Disease Studies and in MSHA, Davitt McAteer (the son of a miner) was named assistant secretary and Andrea Hricko deputy assistant secretary. All three had a longstanding interest in dust diseases and the need for the federal government to play an important role in protecting workers' health. Together, they refocused federal attention on silicosis, culminating in the 1997 National Conference to Eliminate Silicosis, which attracted over six hundred federal employees, industry representatives, union officials, and public health workers.

The national conference broadened the discussion of silicosis to include coal miners in addition to oil workers, foundry workers, sandblasters, and hard rock miners. While silicosis had long been considered a danger to hard rock miners in the West, MSHA officials' concern over silicosis stemmed from fundamental changes in the technology and political economy of coal mining in West Virginia, Pennsylvania, and Kentucky. The Appalachian coal fields were once the richest and most easily accessible source of coal in the nation; in recent decades, the veins of coal have become thinner and deeper in the ground, which has meant that miners have had to drill through silica-laden rock to reach the coal. As a result, miners are now endangered not only by coal workers' pneumoconiosis—which has plagued anthracite miners for decades—but also by silicosis (Personal communication, Davitt McAteer, LLB, July 1997).

In a 1994 letter to MSHA, a Pennsylvania insurance company said it was paying claims on young surface miners who had died from silicosis and questioned what MSHA could do to prevent the disease. In response, an MSHA manager, Jack Kuzar, spearheaded a joint MSHA/NIOSH X-ray screening and outreach program in Johnstown, Pennsylvania, to investigate the magnitude of the problem and to educate coal miners and mine operators on ways to reduce exposure to silica-containing dust. Eight of 150 miners screened had silicosis (Personal communication, Andrea Hricko, MSHA, July 1997). This spurred MSHA to make silicosis one of its top priorities, in addition to its longstanding concern over coal workers' pneumoconiosis.

Prospects for the Future

Today, scientists, policy makers, industrial hygienists, labor unions, and industry representatives are reassessing the danger that silica sand poses to the health of an estimated 2 million workers. The past three years have seen two international conferences evaluating the scientific evidence of the link between silica and cancer, an International Agency for Research on Cancer (IARC) decision to name silica as a known human carcinogen, a decision by the American National Standards Institute to recommend a ban on the use of sand in indoor abrasive blasting, and the initiation by OSHA and MSHA of a national campaign to eliminate silicosis.

The silica standard is likely to once again become an issue. Although the arguments will be cast primarily as epidemiological and technical debates, the historical record suggests that competing interests play a role in framing the question of when and in what amounts silica is safe or dangerous. In the 1930s, with a severe liability crisis forcing industry to act, standards were established that reflected the economic interests and technical capabilities of equipment manufacturers, industries that used sand, and insurance companies that paid liability claims. During the postwar years, the generally conservative political environment, business efforts to downplay the seriousness of the silica hazard, and the incorporation of silicosis into state workers' compensation laws led to the end of efforts to revise the silica standard. Despite studies indicating that silicosis remained a problem, few voices called for legislative action. In the 1970s, following the passage of the Occupational Safety and Health Act, the issue of the standard once again became important. Only concerted industry efforts and the conservative triumph of the 1980s forestalled the banning of sand as an abrasive in blasting.

With the reawakened attention to silicosis and the flood of lawsuits in Texas, Louisiana, Florida, Mississippi, and other states, industry is faced with a quandary of which way to proceed. On one hand, the American National Standards Institute (a voluntary association of industrial hygienists and industry representatives that has established consensus standards for numerous substances) has called for banning the use of sand in indoor abrasive blasting. On the other hand, history is repeating itself in the formation of the Silica Coalition, "a diverse coalition of trade associations and companies involved in the mining, processing, production, and use of silica and silica containing materials," established in 1997 in anticipation of "OSHA rulemaking to control worker exposure to crystalline silica dust in the not-too-distant future" ("New Coalition" 1997). While the organization is ostensibly aimed at providing "sound science" and legal resources to companies potentially affected by any change in government regulation of silica, it is also clear that increased awareness of the dangers of silica and the resulting threat of litigation hangs over the heads of industry executives.

At a 1997 meeting of companies interested in the silicosis issue, Jean McHarg, a Washington, D.C., attorney, noted that "approximately 2,000,000 workers are exposed to respirable silica annually" and that this posed an enormous litigation problem. "If only 10 percent of occupationally exposed workers (or their heirs) believed their lung cancer is due to their occupational silica exposure," then there was a potential for enormous claims (McHarg 1997).

Had a Democratic administration come to power in 2005 we could have expected that MSHA, NIOSH, and OSHA, along with private industry, would once again face a moment of decision. The cooperation between government and industry that led to a national conference in March 1997, in marked contrast to events of the past fifty years, created enormous attention and activity around the issue of silicosis. For the past few years under a Republican administration the issue at the federal regulatory level has been muted, with the focus shifting from regulation to liability suits. The judiciary has once again replaced the legislature and regulatory agencies in adjudicating responsibility for the damages to workers. NIOSH still recommends reducing the permissible exposure limit by half and banning the use of sand in abrasive blasting. If a new administration reestablishes the important role that OSHA once played, we may find that attempts to reduce the permissible exposure limits, for example, would pit the regulators in government against the regulated in industry. Similarly, attempts to ban sand as an abrasive, one of the longest standing goals of NIOSH, will antagonize both providers of sand and end users, despite the availability of cost-effective alternatives to sand for abrasive blasting. As the dangers of silica are more widely publicized, more diseased workers are likely to file lawsuits. Recent attention to the carcinogenic effects of silica and the resulting fear that silica is not only an occupational, but also an environmental hazard will undoubtedly broaden the debate and generate new conflicts. For much of this century the silica issue has arisen in the context of toxic tort cases that forced public health officials and federal and some state administrators to acknowledge the devastation caused by this disease. It is not surprising, therefore, that recent neoconservative legislators have sought to limit the ability of workers and consumers to bring these suits to court through tort "reform" legislation.

Notes

A portion of this chapter previously appeared in *Public Health Reports* 113 (July–August 1998).

1. The names and some identifying information have been changed in these vignettes.

References

Blair A. 1974. *Abrasive Blasting Respiratory Protective Practices.* Cincinnati, OH: National Institute for Occupational Safety and Health, Division of Laboratories and Criterion Development. April.

Glater, J. 2003. "Suits on Silica Being Compared to Asbestos Cases." *New York Times,* September 6.

Goodier, J.L., E. Boudreau, G. Coletta, and R. Lucus. 1974. *Industrial Health and Safety Criteria for Abrasive Blast Cleaning Operations.* Cambridge, MA: Arthur D. Little, September. Contract No.: HSM 99–72–83. Sponsored by NIOSH Division of Laboratories and Criteria Development.

Markowitz, Gerald, and David Rosner. 1995. "The Limits of Thresholds: Silica and the Politics of Science, 1935 to 1990." *American Journal of Public Health* 85: 253–62.

McHarg, J. 1997. "Toxic Tort Litigation Overview." In *Proceedings of Silica in the Next Century: The Need for Sound Public Policy, Research and Liability Prevention Efforts,* ed. P. Boggs. Washington, DC: March 24.
"More on Proposed OSHA Standard for Crystalline Silica Use." 1975. *Materials Performance* (May): 41–42.
National Institute for Occupational Safety and Health. 1974. "Criteria Document: Recommendation for a Crystalline Silica Standard." *Federal Register* 39: 250 (29 CFZ Part 1910).
"New Coalition Will Seek 'Sound Science' on Silica Health Hazards." 1997. *Inside OSHA* (June 30): 1.
"Report of Preliminary Meeting." 1975. In Silica Safety Association Exhibit, vol. 8: 222–23.
Rosner David, and Gerald Markowitz. 1991. *Deadly Dust: Silicosis and the Politics of Occupational Disease in Twentieth-century America.* Princeton, NJ: Princeton University Press, February 5.
Silica Safety Association Newsletter. 1982. In Silica Safety Association Exhibit.
"Sline to Dear Sir." 1975. In Silica Safety Association Exhibit, March 21. vol. 1: 49.
Unpublished data [in SSA]. 1977. Courtney and Company, Deer Park, Texas, February.
Wright, B.C. 1977. "Future of Abrasive Blasting." *Proceedings of the Regional Meeting of the National Association of Corrosion Engineers.* October 5.
———. 1981. "Silica Safety Update: History, Current Endeavors, Future Plans." In Silica Safety Association Exhibit, May 26. vol. 9:84.
———. 1983. *Silica Safety Association Newsletter.* In Silica Safety Association Exhibit. June 22.
Ziskind, M. 1974. "Accelerated Silicosis in Sandblasters: Terminal Progress Report, June 1, 1971–August 31, 1974." New Orleans: Tulane University School of Medicine; Contract No. 5 ROI CH 00387. Sponsored by National Institute for Occupational Safety and Health.

5

How Safe Are U.S. Workplaces for Spanish Speaking Workers?

Laura H. Rhodes

This chapter will explore the plight of Spanish speaking workers in the neoliberal, American workplace. Hispanics experience low wages and work in tough jobs, and it does not look like there is an end to this in sight. Black and Hispanic workers also experience poverty at much higher rates than whites (BLS 2002a). According to immigration data, by 2008 the Hispanic labor force is projected to be larger than the black labor force, primarily because of faster population growth (NAS 1971).

In 2002, the United States Bureau of Labor Statistics (BLS) reported that 840 Hispanic workers were fatally injured at work. Hispanic worker fatalities accounted for 15 percent of the 5,524 total fatal work injuries that occurred in the United States in 2002. The rate of five fatalities per 100,000 workers recorded for Hispanics was 25 percent higher than the rate of four fatalities per 100,000 recorded for all workers. Fatal injuries to Hispanic workers have been on the rise since 1992, when they recorded a low of 533. Moreover, the upward trend in Hispanic worker fatalities has been driven by an increasing number of fatalities to foreign-born Hispanic workers, who in 1992 accounted for 275 fatilities or 52 percent of fatalities to all Hispanic workers, and in 2002 accounted for 577 fatalities, or 69 percent.

These data underreport the real magnitude of the problem because they do not include the illegal or undocumented Latino immigrants in the United States. Hispanics also have lower access to health services, are underrepresented in clinical trials, and live in more polluted communities than their white counterparts. Moreover, these symptoms of environmental injustice and health disparity are more acute at the U.S.-Mexico border, and are prevalent throughout Latin America and the Caribbean (EOSH 2004).

The discussion of an increase in the labor pool in lower paying jobs is moving to front-page news as President Bush's proposed "guest worker plan" has its proponents and its opposition. For example, the UNITE HERE Union reports that the plan would mean a gain of several protections including coverage by the U.S. minimum wage and workplace safety rules, retirement benefits, ability to open bank accounts, and freedom to travel between the United States and their home countries. The drawback is that this is a limited time opportunity with perhaps one chance to renew their status. Many illegal workers are not interested in participating in such a plan because they want to stay in the United States despite what most would consider long-term low wages (Zerembo 2004).

The safety of Hispanic workers in the U.S. workplace within the neoliberal economic environment is a worthy inquiry as their conditions are easily linked to governmental policy. Cutting the Occupational Safety and Health Administration's (OSHA) budget, or not increasing it to meet labor exposures, means fewer inspections of hazardous workplaces and is just one example. Furthermore, the privatization of inspection activities, as proposed by Senator Michael Enzi's (R-WY) bill, presented in the name of efficiency, likely will not provide safer work sites. Neoliberalism is also characterized by a deregulated workplace; for example, the repeal of OSHA's ergonomics standard.

Occupations of Many Spanish Speaking Workers

Hispanics are employed in all sectors of the economy. Occupations where many Hispanics are employed include construction, landscaping, hospitality (hotels/motel), clothing manufacturers, restaurants, and agriculture. This chapter will focus on these sectors, as they are highlighted by BLS and OSHA as troublesome and BLS data support that a significant percentage of Hispanics are employed in these occupations. OSHA also reports high employee injury rates in these occupations and/or the potential for numerous risk factors that may contribute to the injuries. The extent of the problem for immigrant workers is not known. In order to get a better grasp of the true picture, a recent UCLA Labor Occupational Safety and Health (LOSH) program research project identified these industries as primarily employing Hispanic workers by conducting a study of California workers (95 percent of whom were not U.S. natives) (Brown 2002).

Many of these occupations are low paying and demand hard physical work that is associated with a higher injury rate. American-born white workers are reluctant to take these jobs because of the imbalance between the wage scale versus the physical demands and environmental conditions of the occupation. According to Joe Reina, deputy regional administrator for OSHA Region 6, "with some exceptions, Hispanic immigrants in large numbers tend to seek jobs involving unskilled manual labor" (Parker 2005, 34). One of the driving forces behind President Bush's recent immigration proposal is to fill the low-paying jobs in the aforementioned occupations that Americans do not want. "Anglo America has abandoned the construction industry. African Americans have abandoned it, too," says Senaido Adam Travino, chairman of the board for the Hispanic Contractors' Association of Dallas/Ft. Worth, Texas (Halverson 2003). The Guest Worker Plan intent, in part, is to build the labor pool in the neoliberal workplace.

The author will explore the tasks, physical demands, and environmental conditions of these occupations. The *Occupational Outlook Handbook,* published by the BLS, is a nationally recognized source of career information, designed to provide valuable assistance to individuals making decisions about their future work. Revised every two years, the handbook describes what workers do on the job, working conditions, the training and education needed, earnings, and expected job prospects in a wide range of occupations. The *Occupational Outlook Handbook* was used to determine the knowledge, skills, and abilities required for the select occupations that will be explored in depth in this chapter (see Table 5.1).

All of these jobs involve tasks that are repetitive in nature, demand hard physical labor, and present challenging work environments to the employees. The combination of these factors leads to high rates of injury and illness in these industries and will be discussed in the next section.

Injury and Illness Incidence Rates

The BLS annually reports the number of workplace injuries, illnesses, and fatalities. Such information is useful in identifying industries with high rates or large numbers of injuries, illnesses, and fatalities both nationwide and separately for those states participating in this program.

Since 1972 the survey has reported annually on the number of workplace injuries and illnesses in private industry and the frequency of those incidents. With the 1992 survey, the BLS began collecting additional information on the more seriously injured or ill workers in the form of worker and case characteristics. At that time, the BLS also initiated a separate Census of Fatal Occupational Injuries to count these tragic events more effectively than had been possible in the survey (U.S. Dol 2005b).

The Standard Industrial Code (SIC) is a four-digit number assigned to an

Table 5.1

Select Occupational Requirements

Occupation	Job tasks	Physical demands	Environment
Construction laborer	Digs, spreads, and levels dirt and gravel, using pick and shovel. Lifts, carries, and uses building materials, tools, and supplies	Physically demanding work. They may lift and carry heavy objects, and stoop, kneel, crouch, or crawl in awkward positions. Some work at great heights, or outdoors in all weather conditions. Some jobs expose workers to harmful materials or chemicals, fumes, odors, loud noise, or dangerous machinery	Some work at great heights, or outdoors in all weather conditions
Landscaper/ grounds maintenance workers	Dig trenches; apply fertilizers, pesticides, and other chemicals and operate power lawnmowers, chain saws, and power clippers	Many beginning jobs have low earnings and are physically demanding and repetitive, involving much bending, lifting, and shoveling	Most of the work is performed outdoors in all kinds of weather, can be physically demanding
Cleaner, housekeeping	Cleans rooms, carries linens, makes beds, and uses various equipment, tools, and cleaning materials	Spend most of their time on their feet, sometimes lifting or pushing heavy furniture or equipment	Building cleaning workers usually work inside heated, well-lighted buildings. However, they sometimes work outdoors, sweeping walkways or shoveling snow
Sewing machine operator	Sewing functions: assemble or finish clothes or other goods	Operators often sit for long periods and lean over machines. Another concern for workers is injuries caused by repetitive motion	Sewing areas may be noisy. Many older factories are cluttered, hot, and poorly lit and ventilated

(continued)

Table 5.1 *(continued)*

Occupation	Job tasks	Physical demands	Environment
Agriculture worker	Harvests crops using appropriate method, such as picking, pulling, and cutting	Physical in nature and may require much bending, stooping, and lifting	Outdoors in all kinds of weather. Workers may lack adequate sanitation facilities while working in the field, and their drinking water may be limited
Restaurant worker	Perform routine, repetitive tasks such as readying ingredients for complex dishes, slicing and dicing vegetables, and assembling salads and cold items	Workers usually must withstand the pressure and strain of standing for hours at a time, lifting heavy pots and kettles, and working near hot ovens and grills	Working near hot ovens and grills

Source: Bureau of Labor Statistics, *Occupational Outlook Handbook, 2002.*

industry. An industry consists of groups of establishments primarily engaged in producing or handling the same product or group of products or in rendering the same services. Industry definitions used in BLS programs are from the 1987 *Standard Industrial Classification* manual. Because the manual is used by many other federal government statistical programs, it is possible for users to assemble a comprehensive statistical picture of an industry. We will examine the injury rates of our six selected SICs employing Hispanics.

Table 5.2 reviews the injury and illness incidence rates associated with these six occupations according to the BLS. OSHA reports in a February 27, 2004, news release that nationwide the average U.S. workplace had fewer than three cases involving days away and restricted time (DART) injuries for every 100 workers. In four of the industries (agriculture, construction, hotels/motels, and landscaping) the DART exceeds the national level. One concern of the author is that these rates may not reflect the true picture. BLS data certainly have become better over the past decades and have become more important to the safety profession for making meaningful upgrades to worker safety, but employers may not know how to record incidents or may simply choose, illegally, not to report OSHA recordable injuries and illnesses. Also, many employees are simply afraid to tell their employer that they have been injured or have become sick due to work exposures.

Table 5.2

Incident Rates of Nonfatal Injuries by Industry

Industry	SIC	TRC (Total)	DART
Agriculture	01	6.2	3.5
Apparel and other textile products	23	4.6	2.7
General building construction	15	6.2	3.2
Hotels and motels	701	6.7	3.4
Restaurants	58	4.6	1.6
Landscaping	078	5.9	3.3

Source: Bureau of Labor Statistics, 2002.

Notes: The BLS is changing from the SIC to the North American Industry Classification System (NAICS). Developed in cooperation with Canada and Mexico, the NAICS represents one of the most profound changes for statistical programs focusing on emerging economic activities. Establishments that do similar things in similar ways are classified together. NAICS provides a new tool that ensures that economic statistics reflect our nation's changing economy.

UCLA's LOSH program (BLS 2002b) reports that Hispanic workers are reluctant to report injuries and illnesses to their employer. In addition, the BLS only surveys about 80,000 employers annually to compile these data. As a result, OSHA recordkeeping is not foolproof. Admittedly, it has improved over the past decades and the data are beginning to give a "truer" picture of accident/illness experience, but it is important to realize that not all employers know how to report; and particularly important to this discussion is that undocumented workers who are injured would never appear on an OSHA log of injuries and illnesses. So if the numbers seem acceptable in some industries, one must consider if the whole puzzle is in place.

Although the garment and restaurant industries do not exceed the national DART, OSHA has identified garment factories and restaurants as environments with significant potential for uncontrolled hazards. Industry-specific "eTools"—"stand-alone," interactive, Web-based training tools on occupational safety and health topics have been developed by OSHA. These are highly illustrated and utilize graphic menus. Some also use expert system modules, which enable the user to answer questions and receive reliable advice on how OSHA regulations apply to the work site. Some eTools, such as Sewing, appear in Spanish.

The United States has more than 850,000 restaurant industry establishments. On a typical day, four out of ten U.S. adults are restaurant patrons (Filiaggi and Courtney 2003). Typical disabling employee injuries encountered in restaurants include falls, cuts, overexertion, and burns. Estimates of restaurant industry employment vary from 8 million workers, using more restrictive BLS data on SIC 58 (Eating and Drinking Places), to 11.6 million workers, using a more expansive restaurant industry definition preferred by the industry itself (OSHA 2005).

How Are BLS Incident Rates Calculated and What Do They Mean?

Because a specific number of workers and a specific period of time are involved, these rates can help identify problems in the workplace and/or progress an organization may have made in preventing work-related injuries and illnesses. An employer can quickly and easily compute an occupational injury and illness incidence rate for all recordable cases or for cases that involved days away from work. Below are the calculations for total recordable cases (TRC) and DART incident rates.

$$\text{TRC} = \frac{\text{Total number of injuries and illnesses x 200,000}}{\text{Number of hours worked by all employees}}$$

$$\text{DART} = \frac{\text{Number of entries in column H and I of OSHA 300 x 200,000}}{\text{Number of hours worked by all employees}}$$

These rates become important on several fronts. The BLS conducts annual surveys of employers and obtains the data to determine rates for each SIC. OSHA uses these data to develop safety and health programming priorities in order to target industries it identifies as high hazard. This could include development of support materials such as eTools, training, and outreach, as well as the creation of targeted inspection programs. For the employer, these rates can be especially helpful in benchmarking with other organizations. The employer can calculate its rate and then compare itself to the national average. Also, an employer should calculate the rates and then set goals and objectives to continuously improve those rates over time. In addition to OSHA and an individual company's safety efforts, some insurance companies use this information when making underwriting decisions related to workers' compensation premiums.

Now that we know the data show significant losses to workers (Figure 5.2) and that these numbers only scratch the surface since many employees do not record or calculate their TRC and DART incident rates due to the magnitude of the injury and illness rates in our six select occupations employing Hispanics, it is necessary to determine the hazards that have brought them to these levels.

Hazards That Contribute to Injuries and Illness and the Role of OSHA

On an annual basis, the BLS, with OSHA, publishes the most frequently cited federal or state OSHA standards for a specified 2-, 3-, or 4-digit SIC. For many years this was available in print form and could be requested from federal OSHA. Now, the OSHA Web page provides these data and much more that a motivated employer could use to focus its safety and health efforts.

The *top five most frequently cited standards* for each of the six select occupations employing Hispanics will be identified. There are several similarities and these will be discussed in detail. All have unique and significant hazards with varied solutions.

Four of the six occupations have been most frequently cited for hazard communication. In order to ensure chemical safety in the workplace, information must be available about the identities and hazards of the chemicals. OSHA announced a two-part initiative addressing this standard on March 16, 2004. It will be focusing on providing compliance assistance to employers and increased enforcement of the standard. OSHA's hazard communication standard (HCS) requires that chemical manufacturers and importers evaluate the hazards of the chemicals they produce or import and prepare labels and material safety data sheets (MSDSs) to convey the hazard information to their downstream customers. All employers with hazardous chemicals in their workplaces must have labels and MSDSs for their exposed workers and train them to handle the chemicals appropriately.

This particular standard must be a challenge for employers where English is a second language or there are extremely low literacy rates. Under this standard, employers must be certain (document) that employees can *read and understand* hazardous chemical labels and MSDS. But a national adult literacy survey conducted by the Educational Testing Service for the U.S. Department of Education revealed staggering results about the state of literacy in America (ALC n.d.). The study found that 40 to 44 million people functioned at the lowest level, allowing them to perform tasks with "brief, uncomplicated text" (Rhodes and Rhodes 2002). This presents a serious problem for employers when the typical job requires a ninth to twelfth grade reading level, the MSDSs being the best example. More staggering perhaps is that *Safety + Health,* a journal of the National Safety Council, reported that "25 percent of the fatalities OSHA investigates are related to language or cultural barriers" (Parker 2005, 35).

Acacia Aguirre, medical director for Circadian Technologies, Lexington, Massachusetts, an organization that conducted studies on the extended work hours of Hispanics, told *Safety + Health* that "in the long run what should happen is that Hispanics learn English. Some companies are moving in that direction. They are offering English classes in the workplace to their employees" (Naso 2004, 48). Perhaps it is best to consider training supervisors in Spanish, translating documents, and using pictures to communicate workplace hazards as a short-term fix. Teaching workers English, both spoken and written, has a greater societal impact by having a literate population.

According to OSHA approximately 350 electrical-related fatalities occur annually (OSHA n.d.). Sectors where Hispanic workers are concentrated experienced a high frequency of electrical-related citations. Electrical standards ranked among the top five most frequently cited in five of the six sectors. Misuse of electrical equipment can be a source of injury, fire, or even fatality.

Other frequently cited standards among these occupations include means of egress, stairways, and walking working surfaces (see Table 5.3 on page 82). The apparel industry's fourth most frequently cited standard was means of egress. Unfortunately, fires and associated loss of life continue to grip the nation despite well-publicized tragedies over the last hundred years. Life safety efforts got a boost, initially, as the result of the 1911 New York City Triangle Shirtwaist Factory fire when 146 immigrant workers lost their lives due to obstructed exits, locked doors, and inadequate fire escapes. Many immigrant men and young girls jumped to their death in this horrific fire (Kheel Center for Labor-Management 2003). The fire on September 3, 1991, at Imperial Foods in Hamlet, North Carolina, occurred under similar circumstances that led to the death of twenty-five workers and injury to forty-nine others who were unable to exit the building due to locked doors. A highly publicized hotel fire occurred in January 2004 in Greenville, South Carolina, that injured twenty and killed six; this particular fire was ruled an act of arson. A lawsuit against the hotel alleges the hotel owners failed to provide adequate security and directions to exits as well as sprinklers. The suit further alleges that the defendants turned off devices designed to alert occupants to fire and smoke. The hotel was not required to have sprinklers because it had been built in the 1980s. Those who were injured and killed were guests, but it is obvious that if patrons were exposed to these inadequacies, workers, could also be at risk (Paras 2004).

Workplace fires continue to occur and doorways and exits continue to be obstructed in the American workplace despite local building code enforcement and that of OSHA. Normally, a workplace must have at lease two exit routes to permit prompt evacuation of employees and other occupants during an emergency. More exits may be required. Many Hispanic workers in the garment industry endure working conditions criticized by organized labor and advocacy groups.

Machine and mechanical transmission guarding appear four times on the list of top cited standards (Table 5.3). According to OSHA, employee exposure to unguarded or inadequately guarded machines is prevalent in many workplaces. Consequently, workers who operate and maintain machinery suffer approximately 18,000 amputations, lacerations, crushing injuries, abrasions, and more than 800 deaths per year. Amputation is one of the most severe and crippling types of injuries in the occupational workplace and often results in permanent disability. All machines consist of three fundamental areas; the *point of operation,* the *power transmission device,* and the *operating controls.* Despite all machines having the same basic components, their safeguarding needs widely differ due to varying physical characteristics and operator involvement.

OSHA has prepared some specific activities to address the hazards found in these industries, but several of the actions were taken some time ago. More needs to be done more often to tackle these obvious hazards.

Each of these sectors has its own unique hazards and as a result OSHA has specific control measures for some of the industries. We will explore those here.

OSHA's workplace ergonomics standard was issued in November 2000 during the Clinton administration and repealed in March 2001 by the Bush administration by a never-before-used congressional act. This standard focused on worker protection from stressors that create musculoskeletal disorders such as carpal tunnel syndrome and tendonitis due to repetitive, forceful, and awkward work habits. The standard was repealed even though musculoskeletal disorders continue to plague the workforce. The elimination of this standard means that no employer, neither in general industry nor in construction, can be cited for a *specific* ergonomics violation. There still exists a potential to be cited under the General Duty Clause of the Occupational Safety and Health (OSH) Act, but the likelihood of that happening is very low for most employers. Instead of OSHA enforcement addressing this hazard, OSHA has developed technical assistance for employers in the form of Ergonomics eTools, which provide guidance information for workers involved in sewing activities, such as manufacturing garments and shoes, who may be at risk of developing musculoskeletal disorders (MSDs). The eTool is also available in a Spanish version. OSHA also has developed a Construction eTool that will help assist employers to identify and control the hazards that commonly cause the most serious construction injuries, also available in a Spanish version.

Table 5.3

Most Frequently Cited OSHA Violations by Industry

Business/rank	Violation
Agriculture	
1	Hazard communication
2	Electrical, wiring methods, components and equipment
3	General Duty Clause (Section of OSH Act)
4	Guarding of farm field equipment, cotton gins
5	Mechanical power-transmission apparatus
Apparel	
1	Hazard communication
2	Mechanical power-transmission apparatus
3	Electrical, wiring methods, components and equipment
4	Means of egress, general
5	Machines, general requirements
General building construction	
1	General requirements for all types of scaffolding
2	Fall protection scope/applications/definitions
3	Ladders
4	Electric wiring methods, components and equipment, general use
5	Stairways
Hotels and motels	
1	Hazard communication
2	Personal protective equipment, general requirements
3	Blood-borne pathogens
4	Asbestos
5	Electrical systems design, general requirements
Restaurants	
1	Hazard communication
2	Personal protective equipment, general requirements
3	Walking-working surfaces, general requirements
4	Electrical wiring methods, components and equipment
5	Means of egress, general
Lawn and garden care	
1	Hazard communication
2	General duty clause (Section of OSHA act)
3	Pulpwood logging
4	Personal protective equipment, general requirements
5	Occupational head protection

Source: U.S. Occupational Safety and Health Administration.

The administration's more recent cooperative *alliances partnerships* have important loss reduction strategies for OSHA. In order to reduce injuries and illnesses in the lawn and garden industry, for example, OSHA has formed an alliance with the American Lawn Care Association (ALCA). ALCA represents approximately 2,500 professional exterior and interior landscape maintenance, installation, and design/build contracting firms and suppliers nationwide. A team of OSHA and ALCA representatives will meet at least quarterly to track and share information on activities and results in achieving the goals of the alliance. The purpose is to disseminate information and guidance in Spanish and other languages.

The alliance unites ALCA and OSHA in efforts to provide all workers in the landscaping industry, focusing in youth and non–English-speaking workers, with information and guidance that will help foster a safer work environment. According to an OSHA news release (2004a), the alliance will focus on reducing injuries and illnesses associated with manual material handling, motor vehicle crashes, and slip and trip injuries. "This Alliance is an example of how we are striving to reduce injuries in specific industries where the injury and illness rates are too high," said OSHA administrator John Henshaw. "Landscaping is one of seven industries in which we're focusing considerable resources to drastically reduce those rates. Together, OSHA and ALCA can make a positive difference for the lives of thousands of landscape workers and the businesses that employ them" (OSHA 2004a).

"ALCA is dedicated to cooperatively working with OSHA and salutes their willingness to partner with industry," added Debra Holder, ALCA's chief executive officer. "The Alliance will help to further our joint objective of improving safety in the landscape industry through enhanced communication about safety issues, development of safety programs, and an ongoing dialogue and commitment to work together" (OSHA 2003a). ALCA and OSHA will develop and deliver safety training programs for a diverse workforce, using appropriate language, media, and delivery methods. Also, individuals who are bilingual and have knowledge of the landscaping industry will be identified and trained to teach the courses. To reach that goal, both OSHA and ALCA will disseminate information through various media, including their respective Web sites and the creation of electronic assistance tools. The organizations will also participate together in forums and speak or appear at various conferences, such as the Landscape Industry Conference and Green Industry Expo.

In addition to similar, frequently cited standards between these groups, each industry presents its own unique hazards. For example, the hazards seen in landscaping are much different than in apparel. Yet, it is every employer's duty to "provide safe and healthful working conditions" (Section 5[a] [1] of the OSH Act), meaning employers need to communicate those uncontrolled (not engineered out) risks to their employees. Federal OSHA's Web site has many Spanish resources available and Puerto Rico's OSHA has additional documents available for download that an employer can use to disseminate safety and health information to Hispanic employees.

Perhaps the most important part of the hazard identification discussion is the necessity to highlight the safety theory that any time a physical hazard is present (i.e., a citable standard), a management deficiency exists. In other words, if workers are exposed to any hazard the company owners have "dropped the ball." Most safety professionals subscribe to the belief that workers do not *create* hazards but rather management allows them to *arise.* Often, hazards are present because of a lack of engineering controls (management's lack of commitment of capital resources) and administrative controls such as rules or self-inspections (management's lack of assignment of responsibilities, accountabilities, and authority). These data can be quite helpful to professions implementing change. In particular, if an employer does not know where to begin, the most frequently cited standards can be an excellent starting point. Business owners' employee safety and health protections should address each of the areas cited.

Certainly there are states that experience higher Latino populations, but for a state to consider itself immune to the challenges of Hispanic worker injury and illness is naive. California has many Hispanic workers and the protections afforded are dependent on the way occupational safety and health is regulated and fiscally supported. California, like many other states today, is experiencing significant budget problems resulting in fewer CalOSHA inspectors. Workplace safety in California is not enforced by the federal OSHA.

Section 18 of the OSH Act of 1970 encourages states to develop their own OSHA programs, commonly called "plans." Twenty-six states and territories have developed such programs or plans. Although our forefathers perhaps would be pleased with this option of Cooperative Federalism, such an endeavor can be quite costly to the states to run. Each state must create and maintain the plan in such a manner that it is at least as stringent as a federal plan. The cost of technical assistance for the ordinary inspector in the area of safety, industrial hygiene, fire, and ergonomics, not to mention legal matters, can be astronomical. Some states simply would find this financially overwhelming considering states' budget crises.

According to Brown, Domenzain, and Villoria-Siegert (2002), a 2001 report by Nancy Cleeland of the *Los Angeles Times* on labor enforcement found that "by several measures funding, staffing ratios and number of inspections was lower in 2000 than in the last three decades and that the number of CalOSHA inspections had declined by 47 percent between 1980 and 1999."

Brown et al. further report that one month later California's governor cut $3 billion from the budget for labor law enforcement. This fiscal assault is not just occurring in this one state. California has the largest Hispanic population, but other state plans face similar economic circumstances impacting worker protection agencies.

Federal OSHA is planning to conduct 37,700 inspections of workplaces in FY2005, the same number as planned in FY2004 (OSHA 2004b). OSHA has an estimated 1,082 federal inspectors compared to 1,271 in 1990, even though the number of U.S. workers has grown 16.2 percent, according to a recent *USA Today* report (Hopkins 2003). How can so few inspectors make a significant impact on incident rates at millions of legitimate work places? Since the passage of the OSHA budget, a statement was issued (*OSHA Strategic Management Plan for 2003–2008,* 2004) identifying the industries that they will be targeting (U.S. DOL 2005):

- Landscape and horticultural services (SIC 078)
- Oil and gas field services (SIC 138)
- Canned, frozen, and preserved fruits, vegetables, and food specialties (SIC 203)
- Concrete and concrete products (SIC 327, except SIC 3274 and SIC 3275)
- Steel works, blast furnaces, and rolling and finishing mills (SIC 331)
- Ship and boat building and repair (SIC 373)
- Public warehousing and storage (SIC 422)

Hispanic workers may be underrepresented in this list. Certainly, there are sectors listed that employ many of the Hispanic population and the continuing partnership between OSHA and the Hispanic Contractors of America would be expected to make an impact, but the decrease in disabling injuries, illness, and fatalities *will not* be significant for the Hispanic population. It is understood that there are only so many resources available to increase enforcement of the OSHA regulations; only the delusional academic would believe that more tax dollars can be funneled to increase inspections and criminal charges in these sectors.

Regardless of inspections, there continues to be debate over whether OSHA lacks teeth, particularly when an employee is killed in a work-related accident. Senator Jon Corzine (D-NJ) discussed OSHA's lack of prosecution of willful violations, even those that cause death, stating that he wants serious penalties including jail time for those who cause worker death (Hopkins 2003). Corzine has introduced a bill that would amend the OSH Act to significantly increase jail time for wrongful death (from six months to ten years).

At the time of this printing the bill had been referred to the Committee on Health, Education, Labor, and Pensions. The BLS reported that in 2001 workplace fatalities for Hispanics were 20 percent higher than for whites and blacks. Do the families of workers killed receive justice? And perhaps more important, are future deaths prevented by the current policies?

Now that there is an understanding of the specific hazards and OSHA's role in control of those hazards, we will explore the extent of its efforts toward worker protection, specifically, the efforts of federal OSHA over the past several years to prevent or reduce the fatalities, injuries, and illnesses suffered by Hispanic workers.

OSHA's Recent Efforts to Prevent/Reduce Injuries and Illnesses

Despite the published list of focus areas under the most recent OSHA plan, Spanish speaking workers' protection as a whole under U.S. Secretary of Labor Elaine Chao's leadership seems to have moved to the front burner compared to previous secretaries of labor. In other words, the focus areas may not address all the concerns of Spanish speaking workers but it seems that this particular secretary of labor is taking their working conditions and injury and fatality rates more seriously. In 2001 OSHA established its Hispanic Outreach Task Force due to alarming fatality rates published by the BLS, and for several years OSHA has made publications available in Spanish. For example *Todo Sobre la OSHA* (*All About OSHA*) is an excellent booklet describing the history and services related to OSHA (such as National Institute of Occupational Safety and Health [NIOSH] and OSHA state consultation services). Perhaps most importantly it outlines worker protection afforded by the Williams-Steiger OSH Act and the accessibility to enforcement. This booklet also alerts employees that they may not be even covered by OSHA depending where they work.

It is a great stride for OSHA to make these documents available, but because literacy rates are low manuals and training handouts may be worthless. According to Matthew Halverson, author of "Lost in the Translation" (2003), many Spanish speaking employees are reading Spanish at a second grade level.

Because the realization that simply providing translated documents is not the answer to impacting fatality, accident, and illness rates, OSHA has signed agreements with organizations such as trade groups to promote worker safety. One such partnership was announced in March 2002 between OSHA and the Hispanic Contractors of America, Inc. (HCA), which offers resources in Spanish. OSHA also is partnering with the Hispanic Chamber of Commerce, and representatives make regular visits to churches with large Hispanic congregations (Halverson 2003).

One positive move OSHA made in February 2002 was to assist employers with Spanish speaking employees in the development and widespread availability of Spanish training materials, including hazard recognition and controls, via its Spanish Web page. The Web page initially focuses on several areas: an overview of OSHA and its mission, how to file complaints electronically in Spanish, worker and employer rights and responsibilities, and a list of resources for employers and workers. The OSHA Web site, www.osha.gov, also has eTools targeted to this group. For example, La Prevención De Fatalidades focuses on electrical, falls, and trenching fatality prevention. This is a user friendly and informative site completely in Spanish. A second tool (La Costura) targets ergonomics (work-related repetitive stress mostly occurring to the upper extremities and to the back) issues in the sewing industry. This eTool makes workers aware of the potential risk factors for ergonomics-related illnesses. The down side of this tool is that workers, regardless of primary language, rarely have the authority to make necessary workstation changes.

Certainly, these computerized tools are only as good as the availability of Web access compounded by literacy rates. Dependence on employer computer access and access within the workers' communities are the hurdles. Perhaps community cultural centers, hospitals, public libraries, and even churches should promote worker health and safety training through their commitment to providing accessibility to computer terminals with Internet access. The 2002 agreement between the HCA and OSHA specifically stated that OSHA would "work with community and faith-based organizations and other leadership groups to build safety and health awareness within the Hispanic community." Newer OSHA materials focus on the use of photographs to get the messages across in its materials versus concentrating on the actual OSHA standard (Parker 2005).

OSHA needs to continue to upgrade the academic levels of its inspectors. As academically well-prepared inspectors begin to hit the streets, the reputation of OSHA and its effectiveness are likely to improve. Upgrading credentialing has been a priority for the current assistant secretary, John Henshaw. Fortunately, there is also some good news on the financial front that may help move this initiative forward. Recent changes in the law now permit federal agencies to reimburse their employees for expenses related to professional certification, giving additional incentives for those pursuing Certified Safety Professional (CSP) and Certified Industrial Hygienist (CIH) credentials. In years past, government funds could not be used to pay for application or examination fees earmarked for professional certification.

It has been established that significant hazards exist in our six selected occupations where Hispanics are employed, the role of OSHA, and the extent of their success in controlling those hazards. We will now explore why workers continue to seek these hazardous, low-paying jobs and the challenges faced by Hispanics trying to improve their conditions.

Challenges Faced by Hispanics Seeking to Improve Workplace Safety

The Occupational Safety and Health Act of 1970 gives employees the right to file complaints about workplace safety and health hazards. The OSHA complaint process is generally pretty simple and the OSHA Web site makes it easier to overcome the language barriers for Spanish speaking employees. There are different types of complaints employees can file against their employer. A *formal complaint* is one that is written and signed by the employee. The area office must react to a formal complaint. An *informal complaint* can take many forms; for example, a call to the local area office, an unsigned written complaint mailed to an area office, or a fax to an area office. An area office must follow up on such complaints but not necessarily by an onsite inspection. A letter may be sent to the employer indicating an informal complaint has been made and then the employer must respond or use a state OSHA consultation program for assistance in hazard abatement. Most online complaints are addressed by OSHA's phone/fax system. That means they may be resolved informally over the phone with the employer. Written and signed complaints submitted to OSHA area or state plan offices are more likely to result in onsite OSHA inspections.

The OSH Act gives complainants the right to request that their names not be revealed to their employers and OSHA must comply. It has been the author's experience, though, that many employers are well aware of who has filed a complaint, most likely because the employee has been requesting that the employer fix a particular hazard for some time prior to the call to OSHA. Sometimes it is not difficult to deduce which employee wrote the letter. The complainant is covered by OSHA's whistleblower protection legislation that is supposed to ensure that the employer cannot fire, demote, or otherwise punish or treat the employee differently. Most of these types of discrimination complaints fall under the OSH Act, which gives the employee only thirty days to report, unlike other forms of discrimination.

At least one of the following eight criteria must be met for OSHA to conduct an onsite inspection (U.S. DOL n.d.):

1. A written, signed complaint by a current employee or employee representative with enough detail to enable OSHA to determine that a violation or danger likely exists that threatens physical harm or that an *imminent danger* exists. Section 13(a) of the act defines *imminent danger* as "any conditions or practices in any place of employment which are such that a danger exists which could reasonably be expected to cause death or serious physical harm immediately or before the imminence of such danger can be eliminated through the enforcement procedures otherwise provided by this Act";
2. An allegation that physical harm has occurred as a result of the hazard and that it still exists;
3. A report of an *imminent danger;*
4. A complaint about a company in an industry covered by one of OSHA's local or national emphasis programs or a hazard targeted by one of these programs;
5. Inadequate response from an employer who has received information on the hazard through a phone/fax investigation;
6. A complaint against an employer with a past history of egregious, willful, or failure-to-abate OSHA citations within the past three years;
7. Referral from a *whistleblower* investigator; or
8. Complaint at a facility scheduled for or already undergoing an OSHA inspection.

Hispanics accounted for more than half of the country's population growth from 2000 to 2001. Roughly 1.7 million Hispanics either emigrated to or were born in the United States in that year alone, helping this demographic of the workforce grow to more than 14.5 million. (Halverson 2003) The reality for Hispanic workers is that they "shy away from the places that give them the safety information they need" according to Markisan Naso (2004). Workers just do not feel comfortable speaking to OSHA or other government agencies. Perhaps their fears are not unfounded as federal agencies such as OSHA have the ability to report other potentially illegal issues employers are involved in through a referral procedure. For example, if an OSHA inspector believes that there is a child labor or even an immigration problem in a facility, he or she can implement a referral procedure to the appropriate authority. This is outlined in a 1991 OSHA Memorandum of Understanding that states that OSHA's purpose is to:

> set forth a working relationship between the two agencies in order to protect the health and well being of the Nation's work force. It establishes a referral system between the two agencies which will provide for a coordinated

> enforcement effort, thereby securing the highest level of compliance with the Occupational Safety and Health Act and the various Federal Wage and Hour Laws which include: The Fair Labor Standards Act (FLSA); The Davis-Bacon Act (DBA) and related Acts; The Migrant and Seasonal Agricultural Worker Protection Act (MSPA); The Immigration and Nationality Act (the temporary alien agricultural (H-2A) worker provisions); The McNamara-O'Hara Service Contracts Act (SCA); The Walsh-Healey Public Contracts Act (PCA); and The Contract Work Hours and Safety Standards Act (CWHSSA). This MOU is designed to develop cooperation between ESA/WH and OSHA in conducting training programs on the agencies' regulations and requirements and to ensure that ESA/WH and OSHA shall exchange information relating to complaints, inspections or investigations, violations discovered, monetary penalties, and other legal actions to enforce pertinent laws and regulations necessary to ensure coordinated enforcement.

This is not to imply that these interagency agreements should deter complaints or the fulfillment of the law. But the author encourages grassroots efforts to provide safety and health information to workers.

According to Brown, Domenzain, and Villoria-Siegert (2002), 90 percent of workers surveyed were afraid to be injured on the job. In this study workers stated that they feared reporting accidents due to potential employer retaliation. Construction by far is one of the most dangerous occupations for any worker, yet due to economic expansion more than 100,000 construction companies were added from 1990 to 2000 and construction employment rose 30.8 percent to 6.7 million workers. Only the services industries are higher. These positions coupled with a labor shortage give rise to Mexican workers "filling the gap." More than 4 million crossed the border during the 1990s (Hopkins 2003).

According to the General Accounting Office, many Hispanic workers have less than a high school education and often cannot speak or read English. Some are not even proficient in their own language, giving them fewer job options. Many are day laborers and congregate on corners waiting for contractors to offer them cash but no benefits (Hopkins 2003). These conditions prompt workers to accept conditions that would not be tolerated by the average American English speaking worker. Certainly these circumstances contribute to rising death, injury, and illness rates.

How Can the Safety Profession Impact Conditions Affecting Hispanic Workers?

In order to potentially solve this problem or at least impact the rates, this academic has to ask the questions: How can the safety profession impact the conditions affecting Hispanic workers? Specifically, what academic changes need to be made to better prepare the safety professional? And how can the safety professional better serve the Spanish speaking labor force?

Of the five Accreditation Board of Engineering and Technology (ABET) accredited schools offering bachelor's degrees in safety sciences—preparing those who will serve as safety, health, and environmental managers; insurance loss control representatives; risk managers; and the like—none provides course work targeting protection of Hispanic workers. Indiana University of Pennsylvania touches on the subject in Current Issues in Safety (taught by the author). Other schools may be exposing students to the horrors of uncontrolled hazards and lack of information the Spanish speaking worker experiences through multicultural course requirements or "special topics" courses such as at Marshall University. Neither Millersville University of Pennsylvania nor Murray State provides specific courses addressing Hispanic worker protection, which raises the question: Should university professors create courses that focus on such concerns and offer them as seminars, full-academic-load courses, or even as a concentration similar to nursing degrees that focus on populations such as geriatrics?

When surveying students about their interest in such a course, if it counted as a professional elective, one student wrote: "I would not take a course, because I believe that worker protection should cover all races." Another student adeptly pointed to the importance of safety professionals understanding legislative activity such as the Bush position on immigrant workers and their impact on the safety profession when he wrote, "How will Bush's new [proposed] labor law affect [workers' compensation] premiums for employers?" One might surmise that the second student is more cognizant of current issues affecting the professor.

It is shortsighted for schools preparing those who will manage worker safety in a diverse setting to fail to prepare graduates for their safety function. Even at Indiana University of Pennsylvania, introductory Spanish is not an approved professional elective for the safety sciences student. These elective courses and those directed specifically at Spanish speaking worker safety should be encouraged by academic advisers to other majors including human resources management, industrial management, and even engineering.

A freshman level course, Introduction to Safety Sciences, was surveyed ($n = 43$) as to the students' exposure to the Spanish language. Introduction to Safety Sciences is available to all majors at the university so the survey consisted of fifteen majors and twenty-eight nonmajors. Some junior-level safety sciences majors enrolled in Principles of Industrial Safety II (primarily an OSHA compliance course) were also surveyed ($n = 31$). Of the freshman majors, only five had taken Spanish in school. None had it at the college level. Twenty-one of the twenty-eight nonmajors had taken Spanish, six at the university level. The data show that nearly two-thirds of the junior safety sciences students had taken a Spanish course, most at the high school level. These data may imply that other majors recognize the need for a language requirement. Of the five ABET schools, none was found to require second language course work.

Only three of the fifteen freshman majors had ever worked with a Spanish speaking worker, while more than half of the juniors had worked with Spanish speaking employees in the following sectors: construction, landscaping, clothing manufacturing, restaurants, and agriculture.

The reality is that most Safety and Human Resources graduates, from Pennsylvania to California, will encounter this underprotected population. In fact, California immigrant workers were asked where they went for advice on health and safety issues and they responded that they consulted with coworkers, immigrant worker advocacy groups, or labor unions. They seldom turned to their employers for assistance (Brown, Domenzain, and Villoria-Siegert 2002). Furthermore, students need exposure to the Hispanic culture. They must understand motivating factors as they develop and help to implement safety initiatives (Halverson 2003). What many safety professionals may not understand is that the work culture in several Latin American countries, especially Mexico, can instill the idea that work should be done quickly and by cutting corners. According to Halverson, workers are taught that they are expendable and that work must be done with limited resources (Halverson 2003). Hispanic workers need to be assured that safety is a priority within our culture.

Safety sciences graduates need to consider working for employers who represent labor's interest. They could positively affect the plight of Spanish speaking employees and professionalize the staff of government groups such as OSHA. The lure of corporate salaries and upward mobility within corporate structures attracts the best and brightest students. Few students have dedicated their lives to a more noble cause than safety science. Students need to be in touch with their higher calling, protecting all workers, and consider nontraditional sectors for careers.

At the very least, academic advisers should be encouraging graduates to consider meaningful employment with high impact potential, such as labor unions and labor advocacy groups, over high-paying corporate safety management positions. As stated best by John L. Henshaw, assistant secretary of labor for Occupational Safety and Health: "There can be no work more rewarding and no job more fulfilling than helping to protect the lives and well-being of the working men and women who keep our nation strong" (OSHA U.S. DOL 2005).

Furthermore, perhaps as OSHA's staff becomes more professionalized, with positions filled by academically trained safety professionals, as well as when legislation finally elevates the safety profession to the likes of nursing, medicine, and engineering, the accident and illness rate will be significantly impacted. It does appear that the current administration recognizes that OSHA needs to have more academically qualified inspectors and regional office administrators. Now the administration needs to be convinced of the need for more inspectors and the legislators the importance of safety professional licensure.

Perhaps it is unrealistic to believe that a university could produce bilingual safety professionals and maybe it is not necessary (Pierce 2003). Certainly it is realistic to ask that those entering the field have a basic communication level, making them more approachable to Hispanic employees. Furthermore, all graduates need to have a better understanding of the Hispanic culture in order to implement effective safety behavior motivators. One option is to consider providing "command Spanish" to students. Unlike ordinary college level language classes, command Spanish eliminates things like verb conjugation and grammar rules and helps participants to memorize short important phrases and general commands such as "don't lift too much," "well done," "good job," and "watch where you are going" (Halverson 2003). In a business era where ethnic diversity is reaching double-digit rates across the country, accredited programs need to respond (Pierce 2003).

Recommendations

1. Hispanics are the largest minority and OSHA should continue to allocate resources and monies toward bilingual training and resources and continue to upgrade the professionalism of its staff. Employers should be consistently and diligently prosecuted for creating conditions that lead to worker deaths. In addition, university-based safety sciences curricula should stress this to undergraduates and graduates with perhaps a Hispanic safety training track, greater availability of programs like "command Spanish," and/or professional electives focusing on cultural differences and solutions.
2. Civic leaders within the Hispanic community need to increase their safety outreach to Spanish speaking citizens. Civic and religious leaders need to make resources available to their communities. Emerging leaders in the Hispanic community can use already established resources such as those created by the March 5, 2003, alliance between HCA and OSHA. The purpose is for safety and health information to be jointly disseminated through community-based

activities, faith-based organizations, and other leadership groups to build safety and health awareness within the Hispanic community; their goal is to reduce construction deaths among Spanish speaking workers. It is recommend that other associations and individuals representing Hispanic workers pursue access to these resources. This is an important step for the Hispanic community since many undocumented workers may not have access to traditional outlets of safety and health information and material and do not feel comfortable seeking information from government agencies (OSHA 2003a).

3. In order for the injury, illness, and death rates to be effectively lowered for Hispanic workers, the language barrier needs to be overcome. This hurdle has many facets. Supervisors and managers cannot speak Spanish and workers can neither speak English nor read Spanish very well. The solutions therefore need to come in several forms. The bettering of the lives of employees as a human resources function dates back to an era when language courses were offered by the company to immigrant workers as well as cooking and homemaking classes for housewives. The school of thought was to make the employee a better, well-rounded, settled worker in order to achieve the greatest production. That way of thinking needs to be rekindled. The harsh reality is that employers, perhaps unfairly, are burdened with both the literacy and language problems. Human resources departments will have to make these skill-enhancement opportunities available for workers and their families to upgrade their language skills. Further, upper management should consider increasing the language skills of supervisors and managers by using something at least as easy as "command Spanish" to create and maintain a quality, healthy workforce. At the very least, supervisors and managers should be able to alert their employees to potential hazards and sudden warnings.
4. Employers who realize that they are contributing to the demise of their workers, regardless of the primary language, and have now seen the light of reason should begin their safety and health efforts with the most frequently cited standards reviewed above. A business culture should be created that first seeks to control the hazards through engineering controls (machine guarding, ventilation, and workstation design, to name a few) followed by administrative controls (work rules, worker training, rotation to reduce repetitive stress and health monitoring, for example). The last resort for employee protection is personal protective equipment (PPE.) Eliminating noise

exposure is the desired action over providing devices such as earplugs for noise or gloves and respirators for chemical exposures. Many Spanish support materials are available on the OSHA Web site and should be implemented. Furthermore, the employer who recognizes that change is necessary (and morally right) needs to understand that employees *are not* necessarily the source of hazards but rather management allows for hazards to exist. Assignment of specific safety, health, fire, ergonomics and environmental responsibilities, accountabilities, and authorities need to be given to every employee—especially managers. Measuring those activities and assuring completion is the hallmark of a profitable safety culture.

References

Accreditation Board for Engineering and Technology (ABET). 2004. *Accredited Applied Science Programs,* March 15. Available at: www.abet.org/accredited_programs/appliedscience/schoolarea.asp

AmericanLiterary Council. n.d. Available at: www.american literary.com

Brown, Marianne P., Alejandra Domenzain, and Nelliana Villoria-Siegert. 2002. *California's Immigrant Workers Speak Up About Health and Safety in the Workplace,* Health and Safety Policy Brief, UCLA-LOSH. December.

Bureau of Labor Statistics (BLS). U.S. Department of Labor. 2002b. *Youths, Blacks, Hispanics Most Likely to Be Working Poor,* April 16. Available at: www.bls.gov/opub/ted/2002/apr/wk3/art02.htm

———. 2003. *Household Data—Annual Averages 2002–2003 Employed Hispanic or Latino Workers.* Available at: ftp://ftp.bls.gov/pub/special. request/lf/aat13.txt.

Environmental and Occupational Safety & Health (EOSH). 2004. Available at: www.geocities.com/hispanic_eosh/, February 4.

Filiaggi, Alfred J., and Theodore K. Courtney. 2003. "Restaurant Hazard: Practice-based Approaches to Disabling Occupational Injuries." *Professional Safety* 48, no. 5 (May).

Halverson, Matthew. 2003. "Lost in the Translation." *Primedia Business Magazines,* June 1. Available at: www.ecmweb.com/ar/electric_lost_translation/.

Hopkins, Jim. 2003. "Fatality Rates Increase For Hispanic Workers." *USA Today,* March 13.

Keen, J., and J. Drinkard. 2004. "Debate Erupts on Foreign Workers." *USA Today.* January 8.

Kheel Center for Labor-Management Documentation and Archives. 2003. "The Story of the Triangle Factory Fire." Cornell University, April 18. Available at: www.ilr.cornell.edu/trianglefire/narrative1.html.

McFeatters, Ann. 2004. "Bush Offers a New Deal for Illegals." *Pittsburgh Post Gazette,* January 8.

Memorandum of Understanding (MOU) Between the Employment Standards Administration and OSHA. 1991. February 28. CPL 02–00–092-CPL 2.92.

Naso, Markisan. 2004. "Hard Day's Night." *Safety + Health,* National Safety Council, 46–48.

National Academy of Sciences (NAS). 1971. Committee on Occupational Classification and Analysis. *Dictionary of Occupational Titles.* Available at: www.wave.net/upg/immigration/dot.index.html.

OSHA. 2003a. "OSHA and Hispanic Contractors form Alliance: Goal to Reduce Construction Deaths Among Spanish-Speaking Workers." Trade news release, March 5. U.S. Department of Labor.

———. *Strategic Assessment.* 2003b. Situational Assessment Working Group, OSHA IntraNet, U.S. Department of Labor, May.

———. n.d. OSHA Construction eTool: Electrical Incidents. Available at: www.DSWA.gov/SLTC/etools/construction/electrical-incidents/mainpage.html.

———. 2004a. "OSHA Aligns with Associated Landscape Contractors of America." U.S. Department of Labor. Feb. 23. Available at: www.osha.gov/pls/oshaweb/owadisp.show-document?p-table=NEWS_RELEASE&p_id=10687.

———. 2004b. "OSHA Identifies Workplaces with Highest Injury and Illness Rates." U.S. Department of Labor. Feb. 27. Available at: www.osha.gov/pls/oshaweb/owadisp.show_document?p_table=NEWS_RELEASES &p_id=10704 22704.

Paras, Andy. 2004. "Mom Cites No Sprinklers in Hotel Fire Suit." May 24. Available at: www.greenvilleonline.com/news/2004/05/24/2004052432032.htm.

Parker, James. 2005. "Emerging from the Shadows: OSHA Partnerships Give Hispanic Workers a Voice." *Safety + Health* (June): 34–37.

Pierce, David. 2003. "Low English Proficiency and Increased Injury Rates: Causal or Associated?" *Professional Safety* (August): 40–45.

Rhodes, L.H, and D.P. Rhodes. 2002. "Best Hiring Practices Aid Accident Prevention." *Professional Safety* (October): 46–52.

U.S. Department of Labor (U.S. DOL). 2005. Available at: OSHA.gov/sltc/youth/restaurant/index.html. Accessed June 6.

——— 2005b. Injuries, Illnesses, and Fatalities. Available at: www.bls.gov/iif/oshover.htm.

———. n.d. Available at: http://omds.osha.gov/StratPlanIntranet/strategic assessment .html.

———. Construction Outreach Program, "OSHA Act, OSHA Standards, Inspections, Citations, and Penalties." Available at: www.osha.gov/doc/outreachtraining/htmlfiles/intraosha.html.

Vázquez, R. Fernando, and C. Keith Stalnaker. 2004. "Latino Workers in the Construction Industry: Overcoming the Language Barrier Improves Safety." *Professional Safety* (June): 24–28.

Zerembo, Alan. 2004. "Garment Laborers Say Bush Guest-Worker Plan an Ill Fit." *Los Angeles Times,* February 8.

6

Got Air?

The Campaign to Improve Indoor Air Quality at the City University of New York

Joan Greenbaum and *David Kotelchuck*

For more than a century the industrial sector was center stage for safety and health struggles. While industrial issues are still vital, the scope of conflict has broadened to include white collar and service occupations that now constitute three-fourths of the American workforce. In many workplaces, the air quality may be as poor as—or worse than—it is outdoors. Unions are becoming increasingly aware of such issues and the safety and health committee of the Professional Staff Congress of the City University of New York (PSC-CUNY) is taking an innovative approach to the problem with its "Got Air?" campaign.

If and when the dust settles over our struggles for government regulations for environmental health and safety for outdoor ambient air pollution, there remains a massive cloud over the issue of indoor air quality in workplaces. Indeed since three-quarters of jobs in the United States are now in the service sector, the issue of indoor air quality represents a new frontier for labor struggles.

In 1970 the U.S. labor movement and its public health allies fought for and won passage of the first comprehensive federal occupational safety and health law, the Occupational Safety and Health Act (OSH Act), establishing for the first time workplace standards covering most private-sector workers[1] and a federal inspectorate in the U.S. Department of Labor with access to workplaces to enforce these regulations. U.S. industrial unions led the fight for the first wave of OSH Act legislation, which addressed the appalling conditions within industrial workplaces (Levenstein, Wooding, and Rosenberg 2000; Kotelchuck 2000).

Now it is time for another wave of occupational safety and health regulations and enforcement. And this time the emphasis needs to be on the *inside* of offices, schools, hospitals, shopping centers, and the like, where more than 76 percent of the U.S. workforce is employed in usually artificially ventilated indoor spaces (U.S. Bureau of Labor Statistics 2003). And as labor unions have had to adjust to the shifting nature of the workforce and our working conditions, we as health and safety activists within the labor movement also need to focus on the issues that most affect our working conditions—the very air we breathe.

This chapter and the stories it weaves together tell of two often conflicting strands in science and society—the tension between scientifically valid *tests* about environmental conditions and our own *experiences*. In the case studies we present we show that experience and activism combined, particularly within the crucible of union activity, is a potentially powerful way both to remedy immediate problems and to set the agenda for further research. Labor union health and safety struggles, like the environmental movement in general, cannot always wait for tests and scientific interpretations to curb present and indeed potential future harm to our bodies and our working lives.

The Complexity of Air

Addressing indoor air quality (IAQ) issues—now increasingly called indoor environmental quality (IEQ)—is complex and requires a mix of research and activism (see the U.S. Occupational Safety and Health Administration [OSHA] and the National Institute of Occupational Safety and Health [NIOSH] Web sites; see also Godish 2000).[2] Concerns about these issues largely began in the 1970s, especially after the 1976 energy crisis, when architects and engineers designed for new standards that required recirculating building air to save fuel costs in the winter and air conditioning costs in the summer. What has come to be known as "sick-building syndrome" arose in the wake of the 1970s energy crisis with sealed buildings and windows and has contributed to an increased incidence of IEQ problems. Sick-building syndrome is a term used to describe situations where building occupants experience acute health effects, such as headaches and rashes, while at work, but generally find that the symptoms are relieved after a weekend or more spent away from the building.

The increase in the number of indoor air complaints can be seen from the fact that in 1980 only 8 percent of all requests to NIOSH for health hazard evaluations (HHE) were for office environments; in 1990 these had risen to 38 percent of all requests and since then the majority of all HHE requests have been for office environments (NIOSH 1997).

Despite these rising concerns and the widespread recognition that IAQ/IEQ problems are occupational health problems, there is currently no comprehensive OSHA standard for indoor air quality. OSHA standards exist for some individual indoor air pollutants, such as formaldehyde, carbon dioxide, and vapors of many hydrocarbon solvents (U.S. OSHA 2005), but rarely, despite hundreds of indoor air quality investigations in buildings where IEQ complaints such as headaches, dizziness, sleepiness, and fatigue have been reported, do concentrations of these individual air pollutants reach even 1 percent of the OSHA standards (Godish 2000). Calling in OSHA for indoor air sampling is the last thing workers should do, for given the lack of meaningful comprehensive standards management will almost always be told "the air is fine."

The fact that OSHA relies on only individual air pollutant standards suggests to many researchers that there is a need for an understanding of the combined or synergistic effects of toxins in the environment. The phenomenon of synergism is well known to environmental and occupational scientists in studying the ways that the sum of effects is greater than their individual parts. For example, we know from the famous asbestos studies of Dr. Irving J. Selikoff at Mt. Sinai School of Medicine in New York and others that the combined effects of smoking cigarettes and exposure to asbestos dust result in far higher rates of lung cancer than either smoking or asbestos exposure alone (Selikoff, Hammond, and Churg 1968; Benowitz 1997). Handling synergistic effects in general, however, has been a major unsolved problem in public health science, and has come to the fore since the recognition of the importance of synergistic effects in recent decades.

The isolation of individual exposures and the study of their effects have yielded much important knowledge in science for centuries. But given the complexity of mechanisms that govern biochemical phenomena, little progress in predicting or modeling synergistic effects has been made. Currently OSHA and other standards-recommending bodies like the American Conference of Governmental Industrial Hygienists (ACGIH) recommend in the case of multiple chemical exposures that the effects of these exposures be handled additively (U.S. Code of Federal Regulations; ACGIH Worldwide 2003). This at least takes the multiple exposures into account, but conservatively ignores the synergistic effects that in many cases are known to exist. Clearly much research work needs to be done in this area.

In addition to the lack of OSHA standards that take synergistic effects into account, OSHA compounds the problems we face by having no standards for heat stress, cold stress, and indoor humidity conditions. This means that while people working in places that are too hot or too cold know that they feel worse, there are no applicable standards that recognize this. There are indeed problems in adopting such standards—for example, a person's ability to respond to heat stress is significantly affected by his/her prior acclimatization (or lack thereof) to hot working conditions. A good deal of research has been done in these areas and recommendations for how to treat these problems abound. But because of political pressures from industry, OSHA has successfully resisted adoption of any such standard.

Currently the most widely cited and used recommendations are those of ASHRAE, the American Society of Heating, Refrigeration and Air-Conditioning Engineers. ASHRAE has for decades been promoting a series of recommendations on indoor air quality conditions of temperature and humidity and on the numbers of cubic feet of fresh air occupants should be provided in order to have a comfortable indoor working environment (ASHRAE Standard 1992). It recommends indoor temperatures of 72 to 80 degrees Fahrenheit in summer, 68 to 76 degrees in winter, and relative humidities of between 30 and 60 percent throughout the year.

The bottom line for people who work in indoor environments is that they cannot wait until scientists come up with new approaches to synergistic effects and/or OSHA adopts comfort guidelines for indoor air. They are in many cases suffering from headaches, dizziness, breathing difficulties, and fatigue, clearly due to indoor environmental problems. We need to find solutions to these in the here and now.

The PSC/CUNY "Got Air?" Campaign

Fluttering or resting still on the air vents of office and classroom ceilings all over the campuses of the City University of New York (CUNY) are little yellow one-by-five–inch strips with the catchy title "Got Air?" When these are fluttering in the air vents of a room, the occupants—CUNY teachers, staff, and students—know that at least *some* air is flowing into their rooms.

These simple air strips, a kind of people's science if you will, grew out of ideas from members of our union local who joined together as part of what we call the Health and Safety Watchdogs. The strips, for example, can tell us if air is moving in the morning and then be "dead" from say, noon to 2 p.m. when the classrooms are most crowded. While they do not tell our members anything about the quality of the air, they do at least serve as a visible reminder of what one faculty member called "the sleepy-student syndrome." As she put it, the yellow flags called attention to the fact that "it's not my teaching that's putting them to sleep—it's the lack of air."

Our union local, the Professional Staff Congress (PSC) of CUNY, affiliated with the American Federation of Teachers, represents close to 16,000 professional workers throughout the City University of New York, including faculty and staff positions, lab and graduate assistants, full-time employees and part-timers.[3] The university, spread out over the five boroughs of New York City, has hundreds of buildings, most of which are aging and suffering from years of underfunding for operating costs and lack of capital improvements.

The fluttering flags are fruits of a PSC campaign over the past three years. With a new leadership of the PSC elected in 1999, this once sleepy local has become in recent years a powerhouse of union activities. One of these new fronts has been worker health and safety. Since 2000 the union has assembled groups of activists, the Health and Safety Watchdogs, on more than half of CUNY's nineteen campuses. The Watchdogs conduct walkthroughs, surveys, and meetings in buildings around the university. And the number one priority of the Watchdogs has been, and remains, indoor air quality for our members.

Our approach is a bottom-up one, where members report problems to their campus administration and get the local union involved in pushing management to remedy the problems. At Queens College, for example, a large campus sprawling out over a classic quad, the campus-based chapter of the union got the administration to finally close and gut-renovate a large office and classroom building known as Powdermaker Hall. Although faculty and staff had reported problems to the administration for years—such as loose asbestos floor and ceiling tiles, mold, and clogged air vents—the administration's response had always been "there is no money to fix the problems." After the union got involved the building was closed and funds were secured for reconstructing the building from the ground up. And now three years later, union Watchdogs remain on guard in the reopened building. They also hold regular "resident" meetings in different buildings on campus to openly discuss environmental complaints, followed by walkthroughs and inspections where needed. The New York Committee for Occupational Safety and Health (NYCOSH) has provided important technical backup, training, and suggestions for policy recommendations, which the local Watchdogs use to help frame their issues in labor-management meetings.

The management refrain of "not enough funding" used to be the wall against which individual faculty and staff found themselves when making complaints and pointing out building-related problems. Now, however, after two years of NYCOSH-provided training workshops for volunteers from all campuses, the Watchdogs are trained and savvy and management finds itself admitting to problems when the weight of multiple voices, photographs, and evidence is put on the table and, of course, publicized. At John Jay campus in midtown Manhattan, the local chapter of the union put out a newsletter with a photo spread of fire hazard violations and water damage. The newsletter got management's attention and follow-up meetings brought results, not just from local administrators, but from CUNY's central administration, which now put John Jay's North Hall at the top of the list for state-funded renovations.

Since indoor air concerns and associated issues like mold are difficult to adequately test, the strength and visibility of our power to get them remedied lie in the actions of our members within the "shop floors" of their offices and classrooms. Members at Hostos Community College in the Bronx had complained repeatedly of rodent droppings, fleas, and horrendously thick strands of dust hanging from ceilings and air vents. Elevated levels of asthma, statistically more prevalent among the population in the Bronx, were rampant among our members, as were rashes and other respiratory illnesses such as chronic sinusitis and bronchial infections. On an announced Watchdog walkthrough at Hostos Community College, the administration literally spent twenty-four hours sending cleaning and maintenance crews through the buildings in order to make the vents and ceiling tiles look cleaner. Clearly this one-shot management cleanup is not enough but it was a start, and members were pleased over their newfound collective and visible power. With campus administrations now aware of the Watchdogs looking over their shoulders, the union has compiled a list of questions and actions to continue to hold management accountable. Our members have found out that they can ask and expect answers to not very technical, but terribly important questions such as: how frequently are the HVAC filters changed and when were the last routine maintenance and cleanings performed?

Unfortunately as synergistic effects of exposure to pollutants in the air are experienced in the body, it is people and not instruments that are first to detect problems. Thus activism is the forefront of detection, with science as its companion in seeking solutions to the problems. At first many of our members, like workers in other workplaces, were skeptical of their own abilities to identify and solve building-related problems. A history professor voiced perhaps an all too common concern: "I'm not very good at science." For him and many others ordering OSHA/Public Employee Safety and Health or Department of Health tests and calling in the experts were the steps they *thought* they needed to take first in order to document their problems.

But formal testing, as we know from the lack of OSHA standards, is not the first step we want to take. Currently, through training and regular discussions, our members are finding that they can document problems with photographs and through highly visible walkthroughs, where union members walk in groups through a building pointing to problems, writing them down, and asking others to join and show them "eye sores" of concern. Last year our union bought large denim work shirts with an emblem of a Watchdog emblazoned on the back and the words "PSC-CUNY Health and Safety" written all over them. When a group of faculty members begin to walk and talk about problems, others join in and the informal documentation begins. Last winter on a walkthrough of a building on the Brooklyn College campus, extensive ceiling mold was pointed out and photographed. Within a matter of days, management had removed and replaced all effected ceiling tile and insulation, although their overly speedy response would have been improved had we had a chance to order tests about the type and extent of mold infiltration.

At another CUNY college in Brooklyn two indoor air quality problems have arisen during the past two years. In one, spillage of mercury had been taking place over many years in a physics lab as a result of a lab experiment in which students used liquid mercury to measure gas pressure. Following up on concerns of a lab safety officer at the college, a consultant was brought in during the summer of 2001 and found mercury levels in the lab and neighboring areas far in excess of ordinary ambient levels and above indoor levels recommended by the New York City and New York State Health Departments (although still below the legal OSHA limit). Through the union's Watchdog Committee, a fact sheet was developed for PSC members and others at the college and distributed widely. The college responded by moving affected persons to other offices during this summer semester, hired consultants to undertake a cleanup, and agreed to pay for mercury testing at the Occupational Medicine Clinic of the Mt. Sinai Medical Center. In this situation, testing results and the cleanup protocols were shared with the union, and mutually acceptable levels of mercury exposure were achieved. Here then, in a college with an active health and safety committee and local union leadership, the union was able to work cooperatively with the college to achieve health and safety protections for our members.

In the winter of 2003, however, new indoor air quality problems arose at the same college. Poor indoor ventilation in many classrooms had resulted in elevated carbon dioxide levels. In particular, when a union representative made measurements of ambient room levels of carbon dioxide (CO_2) in sixteen classrooms in December 2003, all but one of the classrooms in use at the time of the measurements (this was done in response to and with the permission of the affected faculty) had levels of CO_2 in excess of 1,000 parts per million of CO_2 in the air. These levels are ordinarily considered excessive by ASHRAE. These levels of CO_2 are not considered dangerous in and of themselves—they result from the CO_2 exhaled by the persons in the room—but are an important indicator of inadequate room ventilation. This can result from both faulty ventilation in the room and from overcrowding. Indeed, classroom overcrowding is a common cause of elevated carbon dioxide readings, and represents an important marker for raising both workload and health and safety concerns at the bargaining table. So far management has not rectified the situation, but since safety and health issues are subject to mandatory bargaining we are still negotiating both the workload and the air quality problems.

Many PSC members, like workers in other places, have noted that they have been ridiculed or punished for complaining in the past about problems. More often than not, they were told that "no one else had reported such a concern." For many of our members after years of trying to report problems to management without union backing, the creation of the Watchdogs reminded them that they were no longer isolated "troublemakers." Working together with union activists on their campuses they are able to validate, through ongoing reporting, their own concerns and those of their coworkers. And these concerns, when voiced collectively, can lead to actions at the workplace, as well as movement in collective bargaining. The next section takes up the broader issue of moving environmental issues up to a legislative arena.

Getting Air on a National Scale

The combination of the lack of funds for construction and renovation and shoddy maintenance has provided the circumstances, unfortunately, for the makings of an almost perfect storm over the issue of breathable air in workplaces. Infrastructure, such as the critical but often overlooked workings of the heating, ventilating, and air conditioning (HVAC) systems, has been designed and built to cut costs in the amount of air taken in and recirculated. In addition, cost cutting is also achieved by designing for less interior ductwork. Many engineering plans, for example, place exterior air intake vents for buildings on the ground level, thereby saving on space but leaving them open to pulling in car and truck exhaust. These problems are not unique to the City University of New York. At a recent American Federation of Teachers (AFT) conference on higher education, locals from around the country reported similar problems.

Finally and perhaps most significantly, management has skimped on maintenance and operating costs—in everything from filters through heating and cooling engineering operating shifts throughout the day, evening, and weekends. Many people report that buildings are particularly stuffy or overly hot or cold on Monday mornings as a result of cutbacks in weekend operations. And reuse, retrofitting, and over-occupancy of many buildings, as well as frequent changes in management of the buildings, compound these problems.

The changing nature of work also influences how we experience our time in crowded, poorly designed buildings. As people spend more time sitting in front of computers, their ergonomics problems multiply (Greenbaum 1995; Mogensen 1996). Faculty members, for example, find themselves spending long hours answering student e-mail and writing and printing exams as well as other administrative work. Like other office workers, they find their work process speeding up and their workload more intensified. And the conditions of higher work demands and more limited work controls induce greater stress on the individual, with potentially serious health impacts (Seward 1997).

Within this hot house environment the PSC Health and Safety Watchdogs have joined together with the union's Legislative and Political Action committees to lobby for more statewide funding for our crumbling campuses. These committees, in turn, have joined together with groups from other unions around the state to push for public hearings. For the hearings and our lobbying efforts we put together a photo exhibit of just how terrible our working conditions are. The photos, taken by our members on their campuses and blown up to poster size, have served our members well in getting public attention focused on the need for more funding for new construction and for proper maintenance and repairs. But many next steps remain and chief among them is the need to place indoor environmental concerns on the national political agenda.

The behind-the-scenes cost cutting in construction and maintenance of heating and cooling infrastructures gets brought to the surface in how we feel, act, and indeed perform on the job. Dizziness, headaches, ongoing respiratory ailments, and skin rashes are our visible connection to the invisible world. Just as environmental movements worldwide have used active strategies to bring outdoor air and water pollution to public attention, unions and community and environmental groups are beginning to come together to head off the potential of the perfect storm of deteriorating indoor environmental air quality. And labor's activism, in concert with other environmental groups, can help provide the pressure to gain public funding for much needed research, as well as force managers in both the public and private sectors to pay more serious attention to office and building working conditions on a daily basis.

Notes

1. State and local public employees are excluded from coverage under the federal OSH Act, unless the states in which they work adopt a U.S. Labor Department–

approved plan that must be "at least as effective" as the federal OSH Act (OSH Act, Sec. 18). Such approved state plans must cover all state and local public employees in that state, as well as other workers whom the state also wishes to protect, or to provide a greater level of protection than federal OSHA—for example, the California plan provides greater protection for private agricultural employees than provided for by federal OSHA. As of 2002, twenty-four U.S. states (Arkansas, Arizona, California, Connecticut, Hawaii, Indiana, Iowa, Kentucky, Maryland, Michigan, Minnesota, Nevada, New Jersey, New Mexico, New York, North Carolina, Oregon, South Carolina, Tennessee, Utah, Vermont, Virginia, Washington, Wyoming) and two territories (Puerto Rico and U.S. Virgin Islands) have approved plans and receive federal funds to support these activities. Of these, all cover both private-sector employees and state and local public employees, except New York, New Jersey, and Connecticut, which cover only public employees. See www.osha.gov/fso/osp/index.html for state-plan details.

2. U.S. Occupational Safety and Health Administration (OSHA) and NIOSH bibliographies on this subject at www.osha.gov/SLTC/indoorairquality/index.html, and www.cdc.gov/niosh/iaqpg.html, respectively.

3. PSC-CUNY is Local 2334 of the American Federation of Teachers, AFL-CIO.

References

ACGIH Worldwide. 2003. "Appendix C: Threshold Limit Values for Chemical Mixtures." In *2003 TLVs® and BEIs®*, 72–74. Cincinnati, OH: American Conference of Governmental Industrial Hygienists.

ASHRAE Standard 55–1992. 1992. *Thermal Environmental Conditions for Human Occupancy.* Atlanta, GA: American Society of Heating, Refrigeration and Air-Conditioning Engineers.

Benowitz, Neal. 1997. "Smoking and Occupational Health." In *Occupational and Environmental Medicine.* 2nd ed., ed. Joseph LaDou, ch. 42, 72–74. Stamford, CT: Appleton and Lange.

Godish, Thad. 2000. *Indoor Air and Environmental Quality.* Boca Raton, FL: CRC Press.

Greenbaum, Joan. 1995. *Windows on the Workplace.* New York: Monthly Review Press.

Kotelchuck, David. 2000. "Worker Health and Safety at the Beginning of a New Century." In *Perspectives in Medical Sociology.* 3rd ed., ed. P. Brown, ch. 9. Prospect Heights, IL: Waveland Press.

Levenstein, Charles, John Wooding, and Beth Rosenberg. 2000. "A Social and Political Perspective on the History of Occupational Safety and Health." In *Occupational Health.* 4th ed., ed. Barry S. Levy and David H. Wegman, 28–30. Philadelphia: Lippincott, Williams and Wilkins.

Mogensen, Vernon L. 1996. *Office Politics: Computers, Labor, and the Fight for Safety and Health.* New Brunswick, NJ: Rutgers University Press.

National Institute of Occupational Safety and Health (NIOSH). 1997. NIOSH Facts "Indoor Environmental Quality." Available at: www.cdc.gov/niosh/ieqfs.html.

Selikoff, Irving J., E. Cuyler Hammond, and Jacob Churg. 1968. "Asbestos Exposure, Smoking and Neoplasia." *Journal of the American Medical Association* 204: 106–12.

Seward, James P. 1997. "Occupational Stress." In *Occupational and Environmental Medicine.* 2nd ed., ed. Joseph LaDou, ch. 34. Stamford, CT: Appleton and Lange.
U.S. Bureau of Labor Statistics. 2003. *Household Data Annual Averages,* Table 11.
U.S. Code of Federal Regulations, 29 CFR, Subpart Z: "Occupational Safety and Health Standards," Sec. 1910.1000(d)(2)(i). Available at: www.osha.gov.
U.S. Occupational Safety and Health Administration (USOSHA). 2005 Available at: www.osha.gov/index.html.

7

State or Society?

The Rise and Repeal of OSHA's Ergonomics Standard

Vernon Mogensen

The U.S. economy was buffeted by recession at home and growing trade competition from abroad during the 1970s, making capital accumulation increasingly problematic for corporate interests. In order to restore more favorable conditions for capital accumulation, corporate interests mobilized through their policy-planning network to deregulate the economy. Especially problematic for corporate interests was the Occupational Safety and Health Act (OSH Act) of 1970, which "assured Americans the right to safe and healthful workplaces." In addition to giving workers the right to resist unsafe working conditions, which limited corporate autonomy, it forced corporations to absorb workplace safety and health costs that cut into profit margins. The OSH Act threatened corporate profit margins because laissez-faire economists and policy makers considered the costs of injury and illness to be externalities, not part of the cost of doing business. A hidden subsidy to employers, these social costs were largely borne by workers and the community.

Moreover, as the introduction of new technologies and work processes created additional occupational safety and health hazards, workers and their unions could take action to regulate them under the Occupational Safety and Health Administration's (OSHA) protective umbrella. As in a zero-sum game, employers feared that each new regulation would come at the expense of further limiting capital accumulation and corporate autonomy.

This problem is illustrated by the dramatic rise of work-related musculoskeletal disorders (WMSDs) during the 1980s and 1990s and organized labor's attempts to secure the promulgation of the OSHA ergonomics standard.[1] WMSDs rose with the spread of routinized, repetitive-motion, and computer-automated work in both blue-collar and white-collar sectors of the economy. Since the application of the science of ergonomics can help prevent WMSDs, organized labor and its allies fought a decade-long battle with corporate interests to have OSHA promulgate an ergonomics standard. President Clinton's OSHA issued the ergonomics standard in November 2000, but President George W. Bush and the Republican-controlled Congress responded quickly and repealed it in March 2001. It was the first and only time an OSHA standard had ever been repealed. What took them two decades to achieve lasted just two months.

State or Society? Theory and the Dynamics of the Policy-Making Process

What theories of power account for the rise and repeal of OSHA's ergonomics standard? Much has been written over the past twenty years on the autonomy of state officials from societal pressures,[2] and the case study of the making and unmaking of the ergonomics regulation provides a good opportunity to examine the extent to which the state is autonomous of societal interests when it makes policy. Statist theory holds that state officials engage in state building—the expansion of the national state institutions, policies, and power—which increases their administrative capacity to act. Administrative capacity, in turn, is necessary to maintain and increase political autonomy from societal pressures (Skowronek 1982, 19–31). By defining and emphasizing the role of the state, statist theory stresses that there are clearly discernable boundaries between state and societal institutions. Theda Skocpol, a chief proponent of this approach, says that state structures "are shaped and reshaped not simply in response to socioeconomic changes or dominant-class interests, nor as a direct side-effect of class struggles. Rather they are shaped and reshaped through the struggles of politicians among themselves, struggles that sometimes prompt politicians to mobilize support or to act upon the society or economy in pursuit of political advantages in relation to other politicians" (Skocpol 1980, 200). She adds that "both appointed and elected officials have ideas and organizational and career interests of their own, and they devise and work for policies that will further those ideas and interests, or at least not harm them" (1992, 42).

However, statist theory raises a number of questions for investigation. First, if state building—the expansion of the national state—depends on strong state organizations, then how does it explain the fact that OSHA, a weak state organization, promulgated the ergonomics regulation despite the objections of a broad coalition of corporate interests?[3] Second, if successful policies are developed within the state apparatus by state administrators who are motivated by the desire to maximize their institutional power and positions, then how do statists explain the fact that the proposal for an ergonomics regulation came from organized labor? Third, if state officials engage in building state capacity to increase their autonomy from societal pressure, and the promulgation of the ergonomics regulation amounted to a significant increase in OSHA's state capacity (it was OSHA's largest regulation in terms of workers covered), why would state officials subsequently weaken its newly expanded autonomy by repealing the ergonomics standard? Finally, statist theory places great emphasis on the claim that there are clearly discernable boundaries between state and society. If so, then how does it account for the presence of corporate interests within the state apparatus and governmental dependence on corporate interests?

I argue that statist theory is inadequate to explain the policy process involving the rise and repeal of OSHA's ergonomics regulation. Statists underestimate the role socioeconomic factors play in shaping policy making. State officials make policy, but not under conditions of their own choosing. The ergonomics standard's rise was largely shaped by organized labor, and its fall was fashioned by business interests. The corporate attack on the ergonomics standard is part of a larger campaign to dismantle the New Deal/Great Society social regulatory framework under the guise of reform.

Instrumentalists and structuralists offer alternative explanations that emphasize the vital role of socioeconomic interests in shaping the policy-making process. The instrumentalist approach emphasizes the influential role of the capitalist class in the policy-making process. Ralph Miliband writes that "'the state' . . . does not, as such, exist. What . . . [it] stands for is a number of particular institutions which, together, constitute its reality, and which interact as parts of what may be called the state system" (Miliband 1969, 49). He adds that "the capitalist class, as a class, does not actually 'govern'" (55). Rather, capitalists are "well represented in the political executive and other parts of the state system" (55). This gives them strategic control over the public policy-making process. G. William Domhoff has demonstrated that policy-planning networks are important to corporate interests because they mediate differences between different power blocs within the corporate community and conflicts between the capitalist class, labor, and the state. Members of the policy-planning network circulate between procorporate think tanks, the business sector, and policy-making positions in government. According to Domhoff, the policy-planning network includes foundations, think tanks, research institutes situated at major universities, and "general policy discussion groups, where members of the upper class and corporate community meet with experts from the think tanks and research institutes, journalists, and government officials to discuss policy, ideology, and plans (PIP) concerning major issues facing the country" (Domhoff 1996, 29).

Structuralists argue that the capitalist class is too fragmented and politically divided to respond to crises that may occur and rule as the instrumentalists maintain (Poulantzas 1978). Economic crises, class struggle, or problems of uneven economic development within a society may cause a crisis to occur. Consequently, the state's function is to adjudicate intraclass conflicts and attempt to organize them into a coherent whole to maintain the capitalist system's stability. This organizing role views the state as being relatively autonomous of the dominant classes, which gives it the flexibility to rule against the short-term economic interests of class fractions that threaten the long-term stability of the capitalist system. Under this approach, the goal of public policy is the maintenance of the system of social relations that reproduces the conditions for the accumulation of capital. The state's mediating role is further complicated by the potential for the development of disjunctures caused by the uneven development of the economic, political, and ideological structures of society, which are relatively separate from each other. Protective labor legislation that forces employers to invest in occupational safety and health equipment that might undermine the conditions for capital accumulation would be a disjuncture between the political and ideological levels.

Ergonomics and WMSDs

Ergonomics is the science of redesigning the workplace to meet the safety and health needs of the worker in order to prevent ailments such as WMSDs. WMSDs are musculoskeletal ailments of the lower back, shoulders, neck, arms, wrists, fingers, hands, and nervous system. Ergonomics takes a holistic approach to the relationship between the work environment and human factors such as the worker's muscles, tendons, joints, and nerves. It aims to improve job design in order to minimize monotonous and repetitive tasks, and limit work speed-ups, which may contribute to fatigue and stress. There are over twenty types of WMSDs; the more commonly known types are bursitis, carpal tunnel syndrome, epicondylitis, ganglion, tendonitis, and tenosynovitis. Symptoms of WMSDs include tenderness, swelling, tingling sensations, shooting pain in the wrists, fingers, arms, or elbows, and the loss of hand strength and coordination. In some cases, those with WMSDs can no longer perform simple tasks like lifting their child or opening a bureau drawer (NIOSH 1995).

WMSDs affect a broad array of workers performing repetitive tasks, such as auto assemblers, meatpackers, construction workers, delivery workers, poultry cutters, workers doing manual lifting, and clerks using price scanners (AFL-CIO 1997). WMSDs also plague office workers as video display terminals (VDT) have made it possible to routinize and simplify the clerical work process along industrial lines.

Women are at greater risk of developing WMSDs than men. Due to the glass-ceiling effect, women in service and white collar occupations are more likely than men to languish in jobs that require repetitive motions. A literature review of fifty-six studies on women and WMSDs conducted at the Ohio State University found that they are twice as likely as men to develop upper body WMSDs, and up to eleven times more likely to develop WMSDs than men (Croasmun 2004, 1). This finding supports National Institute of Occupational Safety and Health (NIOSH) studies that found that "women workers are at disproportionately high risk for musculoskeletal injuries on the job, suffering 63 percent of all work-related repetitive motion injuries" (NIOSH 2000).

Large stakes were involved in the battle over the ergonomics standard. WMSDs have been the fastest growing occupational illness in the United States for more than two decades. According to OSHA there are more than 1.8 million incidences of WMSDs in the United States annually; over 600,000 cases are serious enough to cause employees to lose work time. Carpal tunnel syndrome, one of approximately twenty WMSD-related illnesses, is so debilitating that half of all cases result in thirty lost workdays, more than any other occupational illness. OSHA estimates that the economic cost to society of WMSD-related injuries in terms of workers' compensation claims is $20 billion, and that number rises to $54 billion when lost work time and lost productivity are included (NAS 2001; Jeffress 2000, 1).

The Underreporting of WMSDs and Crisis Management

The failure of both the Reagan and Bush administrations to address the growing problem of WMSDs head on allowed it to reach crisis proportions. In 1981, WMSDs accounted for only 18 percent of all reported occupational illnesses, but by 1993 they had mushroomed to 61 percent of all cases, a 34 percent annual growth rate for the ten-year period (BLS 1994, 2).

The U.S. Bureau of Labor Statistics' (BLS) reliance on the good-faith efforts of employers to report injuries and illnesses leads to underreporting, which makes it more difficult for OSHA to monitor trends and evaluate the need for regulation. OSHA relies on BLS figures, which are based on the OSHA Log 200 forms employers file when workers first report injuries to doctors. But BLS figures exclude federal, state, and local government workers, and the growing numbers of the self-employed. Other factors also contribute to the underreporting of WMSDs including nonunion workers who are afraid to report injuries for fear of being fired or who work through the pain or take personal days to recuperate. Another problem arises when doctors either misdiagnose musculoskeletal disorders, or as company doctors may be prone to do, attribute WMSDs to nonwork-related causes. During the lobbying effort to defeat the ergonomics standard, corporate interests erroneously claimed that employers would be forced to pay for workers' musculoskeletal disorders caused by nonwork activities such as playing golf or softball (Harper 2001). As a result, the true number of WMSDs is actually higher than BLS figures indicate. While the BLS puts the number of workers who lose work time due to WMSDs at 600,000, the American Public Health Association (APHA) estimates that more than 775,000 workers suffered WMSDs and overexertion injuries in 1995 (APHA 1996, 1). OSHA also overlooked numerous respected epidemiological studies, some conducted by NIOSH, that demonstrate the need to rely on more than one source of statistics. The underreporting of the data gave business interests a structural balance of advantage in the policy debate since it made it more difficult for organized labor to substantiate the magnitude of the WMSD crisis.

Phase One: The Reagan and Bush Administrations Resist Regulation

As the WMSD problem grew during the early 1980s, organized labor, women's groups, and local committees for safety and health, like the New York Committee for Occupational Safety and Health, banded together to seek safety and health protection. Unions such as the Service Employees International Union (SEIU) led by John Sweeney joined forces with Karen Nussbaum's 9 to 5: National Association of Working Women. They established District 925 to organize and represent women workers—especially those using VDTs—whose workplace and social issues had been neglected by organized labor (Sweeney and Nussbaum 1989, 153ff).

The political campaign for an ergonomics standard went through two strategic phases: phase one during the 1980s and phase two in the 1990s. In phase one, during the early 1980s, organized labor and its allies followed a two-prong strategy. First, they took the federal government route by taking the formal step of asking the Reagan administration's OSHA to issue an ergonomics standard. In May 1984, after more than four years of study, NIOSH recommended to OSHA that it begin work on an ergonomics standard. NIOSH's mission is to conduct medical and scientific studies of occupational safety and health hazards and make recommendations to OSHA for possible action. But OSHA is not required to act on NIOSH's recommendations, and the Reagan administration ordered OSHA to ignore them. In statist parlance, OSHA is a weak state agency lacking the capacity to act autonomously. After two years of inaction, Dr. Roger L. Stephens, OSHA's only ergonomist, lamented that the underfunded and understaffed agency "will never be able to force the needed job redesign or even lead the way" ("Job Redesign" 1986, 1). Rebuffed by the Reagan administration, organized labor pursued the second part of its phase one strategy, a state and local legislative campaign. Faced with massive opposition from the Computer and Business Equipment Manufacturers Association and trade associations representing numerous *Fortune* 500 corporations that used thousands of VDTs, these strategies proved unsuccessful (Mogensen 1996, 127–42).

As WMSD rates in the meatpacking industry reached crisis proportions, the United Food and Commercial Workers, which represented many injured workers, pressed OSHA to address the issue. Resisting regulation, the Reagan administration was forced to resort to the OSH Act's general duty clause, which grants OSHA the power to assess employer penalties for hazardous working conditions not covered by a standard. The Reagan administration instituted a large fines policy in the meatpacking industry, which made headlines but little headway in stemming the rising tide of WMSDs. The Reagan and Bush administrations also ignored warnings by NIOSH that the rapid proliferation of VDTs in white-collar occupations is an "especially important factor" in the rise in WMSDs, and by OSHA that it has "generated new and pervasive sources of biomechanical stress to the musculoskeletal system" (U.S. Congress, House 1989, 118). This problem was compounded by the fact that the BLS did not keep data on VDT workers with WMSDs (Mogensen 1996, 14). Consequently, OSHA administrator Gerard F. Scannel could not give a reliable figure on the number of VDT workers with WMSDs. But the *New York Times,* paraphrasing his words, said "the proliferation of computer technology in many industries was a significant cause of the increase in repetitive motion disorders" (Kilborn 1989).

Phase Two: The Campaign for the Ergonomics Standard

The rise and repeal of the ergonomics standard was bookended by the two Bush administrations. In August 1990, George H.W. Bush's secretary of labor Elizabeth Dole announced that OSHA would begin working on the ergonomics standard. George W. Bush's decision to repeal the ergonomics standard in 2001 is consistent with his family's connections to corporate interests, but why did his father's administration allow the rule-making process to begin?

First, the crisis of WMSDs in meatpacking and other occupations had reached crisis proportions. The general duty clause of the OSH Act was not intended for use on a regular basis to control a specific hazard. Second, unlike 2001 the Democrats controlled Congress in 1990 and used their majorities to hold hearings and press the Bush administration to begin work on the standard. Third, there is usually less organized political opposition at the start of work on a proposed standard than when a final rule nears publication in the *Federal Register*. Finally, the U.S. economy was more integrated into the global economy in 2000, and the corporate pressure on the president and Congress to deregulate labor standards was more intense than it was in 1990.

At first, the Bush administration continued the Reagan administration's policy of inaction. As a weak state agency OSHA lacked the resources to launch an ergonomics program during the Bush administration. Roger Stephens, OSHA's only ergonomist, testified before Congressman Tom Lantos's (D-CA) hearing on the dramatic rise in WMSDs in 1989 that the safety agency needed at least ten (one per region) ergonomists, plus a Washington-based planning and research team, just to start an ergonomics program (House 1989, 110). "We're not happy with OSHA's incredibly slow progress," Margaret Semanario, the AFL-CIO's director of health and safety said. "This does not appear to be a priority at OSHA and we think it should be," she added (House 1989).

In July 1991, the AFL-CIO and thirty affiliated unions criticized the Bush administration's preference for voluntary guidelines, which they feared would not apply to most industries. They pressed Secretary of Labor Lynn Martin to issue an emergency temporary standard immediately. She refused, but labor's lobbying kept the rule-making process alive, and it paid off when the Bush administration finally issued the Advanced Notice for Proposed Rulemaking (ANPR) in June 1992 (AFL-CIO, 2004, 1).

Corporations and Republicans Mobilize Against the Ergonomics Standard

The attack on the ergonomics standard was the corporate community's response to the policy disjuncture created by OSHA's issuance of the ANPR. It was part of a broader campaign to "reform OSHA" that would maximize profit margins and management autonomy by disciplining the labor force and limiting the rights of workers, many of whom are women, unorganized, and working on a contingent basis for low pay. In June 1994, the National Coalition on Ergonomics (NCE), an alliance of more than 300 corporations and trade associations with a $600,000 budget, announced its opposition to the ergonomics standard. The NCE was formed by the U.S. Chamber of Commerce (USCC), which bills itself as "the world's largest business federation," representing more than 3 million businesses nationwide. It has a strong small business political orientation; small businesses with less than 100 employees makeup more than 96 percent of its membership. The USCC's president and chief executive officer, Thomas J. Donohue, was formerly the head of the American Trucking Association, another staunch opponent of the ergonomics standard. Together with the National Association of Manufacturers (NAM), the National Federation of Independent Business (NFIB), the Small Business Survival Committee (representing 40,000 businesses), and the Labor Policy Association (representing 225 businesses), they constituted a formidable lobbying alliance.

The political push to defeat the ergonomics standard started in the U.S. House of Representatives and had its greatest support there. The small business sector is well represented in the House, which, like the Senate, was under Republican Party control from January 1995 through the ergonomics standard's repeal in March 2001. The political influence of small business interests is magnified to a greater extent in the smaller sized congressional districts than in the Senate. Many small-business people have gone on to political careers as representatives. House majority leader Tom DeLay (R-TX) owned a pest control company in Houston that was fined by OSHA. DeLay entered politics to stop what he viewed as the onerous OSHA regulation of small businesses. Known as "Mr. Dereg" in business circles, DeLay quickly attracted corporate campaign contributions and rose up the Republican ranks to become majority whip under House Speaker Newt Gingrich (Maraniss and Weisskopf 1995a). DeLay was the principal architect of the opposition to the ergonomics standard in the House.

Large corporations joined in a coalition with small to midsized businesses to oppose the ergonomics standard. The Washington, D.C.–based Business Roundtable is the most powerful peak association representing the viewpoint of most large corporations in the halls of government. A highly influential policy-planning organization composed of the chief executive officers of 200 large U.S. corporations, it represents a cross section of industries and regions throughout the United States. The Business Roundtable was the leader of the corporate campaign to convince Congress to "reform OSHA," which meant less reliance on OSHA's power to regulate and fine corporations and more voluntary cooperation. It was joined by the innocuously named Center for Office Technology, which was jointly financed by the computer manufacturing industry and large corporate users of VDTs. The Business Roundtable's effort to reform OSHA was part of its larger campaign to deregulate the U.S. economy so that it would be more in sync with the economic requirements of the free-market, global economy. From the corporate perspective, OSHA, environmental, and other protective regulations were seen as barriers to free trade among the world's nations.

Opponents of the ergonomics standard claimed that it would impose burdensome costs on employers; some corporate estimates were as high as $123 billion a year. OSHA countered that annual implementation costs would be closer to $4.2 billion, and would result in savings of $9.1 billion (Jeffress 2000, 5; Congressional Record 2001, S1845, S1849). But whose cost assessments were more accurate? In 1995 the U.S. Office of Technology Assessment (OTA) released a revealing study of the cost of implementing OSHA standards including cotton dust, lead, formaldehyde, and vinyl chloride that had been in effect for more than five years. The OTA concluded that industry consistently overestimated the actual cost of compliance by wide margins, and OSHA's estimates were the most accurate, saying that it "has generally performed this task with workable accuracy—that is, standards determined by OSHA to be 'feasible' in the course of its analytical deliberations have usually proved to be so when industries took the necessary steps to comply" (OTA 1995, 10). Although OTA was a valuable and respected research resource for Congress, the Republicans quickly eliminated it when they gained power in 1995.

Big business also argued that it could not afford the cost of protecting its workers in a globalized free-market economy. In 1993 the Business Roundtable warned the Democrats, who controlled both Congress and the presidency for the first time in twelve years, that "cooperation in our efforts to improve workplace safety and health is critical if we also want to make our workplaces globally competitive" (Business Roundtable 1993, 2). Senator Tim Hutchinson (R-AR) echoed this view during the floor debate on the repeal of the ergonomics standard when he said: "To all of those today who stand on the floor and champion workers' rights, this rule will result without doubt in sending jobs overseas where there are often no worker protections at all" (Senate 2001, S1848). But free-market advocates like Senator Hutchinson are not opposed to outsourcing U.S. jobs overseas and not strong advocates for workers' rights at home. Otherwise, he would have voted against the ergonomics standard's repeal. Moreover, this argument poses more of an ideological threat than an economic certainty. Many of the jobs that are covered, such as manual handling, service, and repetitive motion tasks, are permanent parts of the working fabric of the domestic economy. The function of the globalization argument is to provide political cover for the ideological attack on workers' rights to a safe and healthful workplace.

Small and midsized businesses perceived the ergonomics standard to be more of a threat than big businesses did because they would be less able to absorb the cost of compliance. Big businesses understood that ergonomics not only enhanced employee safety, but maximized morale and productivity as well. Representatives of large corporations came to OSHA's stakeholders meeting in September 1998 in Washington, D.C., to report on how successful their ergonomics programs were at preventing WMSDs (Lichterman 1998).[4] Thirty percent of large U.S. companies have ergonomics programs (Jeffress 2000, 1), but they did not want an ergonomics standard that would limit their control over the workplace.

Corporate Policy-Planning Networks

Policy-planning networks are the means by which elites circulate between procorporate think tanks, the business sector, and the upper echelons of government. G. William Domhoff has shown that policy-planning networks are crucial to corporate interests because they mediate differences between different power blocs within the corporate community and conflicts between the capitalist class, labor, and the state (Domhoff 1996, 29ff). The career paths of three key policy makers involved in efforts to undermine and repeal the ergonomic standard illustrate how the policy-planning network operates.

The three are George W. Bush's top two Labor Department appointees, Labor Secretary Elaine L. Chao and Solicitor General Eugene Scalia, and United Parcel Service (UPS) lobbyist Dorothy Strunk. Elaine Chao, the wife of influential Republican senator Mitch McConnell (KY), served as a Distinguished Fellow at the conservative Heritage Foundation from 1992 to 2001. She was a vice president at American Express Bank and held high-level positions in cabinet departments during the Bush and Reagan administrations. Like President Bush, Chao has a Master of Business Administration from Harvard Business School.

Eugene Scalia, a leading opponent of the ergonomics standard, is the son of conservative U.S. Supreme Court Associate Justice Antonin Scalia. As a partner in the Washington-based corporate law firm of Gibson, Dunn and Crutcher, Scalia lobbied to defeat the OSHA ergonomics rule for clients including UPS, Anheuser-Busch, and the NCE. He was also involved in corporate efforts to defeat the state ergonomics standards in California and Washington. Scalia has been associated with the corporate-funded antiregulatory Cato Institute and the National Legal Center for the Public Interest.

The solicitor general occupies an important position within the Labor Department because it has the responsibility of interpreting and defending labor standards in cases before the U.S. Supreme Court; unlike other agencies it does not have to get clearance from the Justice Department before proceeding. Scalia's appointment was very controversial even when compared to Bush's procorporate appointees. Since Scalia was an aggressive opponent of the ergonomics standard who said it was based on "junk science" and "quackery," labor doubted that he would use his power as solicitor general to defend OSHA's ergonomics program. Fearing that the Democrat-controlled Senate would reject Scalia in January 2002, Bush used his recess appointment power to make him solicitor. However, his recess appointment lasted only until the end of Congress's session in December 2002. Although the Republicans recaptured the Senate in the 2002 elections, they still did not have the votes to confirm Scalia, so in keeping with the revolving door of the policy-planning network, he returned to his corporate law firm.

The third example is Dorothy Strunk, who went from being acting administrator of OSHA during the final months of the Bush administration to becoming a UPS lobbyist on Capitol Hill. UPS, whose delivery and warehouse workers frequently incur WMSDs while manually lifting and carrying heavy packages, has had more safety and health complaints and fines than any other company in OSHA's history. With business operations in every congressional district in the nation, UPS had impressive lobbying leverage on Capitol Hill. It gave House Republicans close to $750,000 during the 1995–96 election cycle in its effort to defeat the proposed ergonomics regulation. Subsequently, Strunk had so much direct access to influential Congress members that she was allowed to help write anti-OSHA legislation. Strunk was so involved in writing Representative Cass Ballenger's (R-NC) 1995 "OSHA reform" bill that an early version was dubbed "Dottie's draft" in her honor. Had it passed, Ballenger's bill would have eliminated the OSH Act's general duty clause. Strongly supported by the USCC and NAM, Ballenger was also chief sponsor of the Small Business OSHA Relief Act of 1996, which was designed to exempt small businesses from OSHA's jurisdiction and would have introduced cost/benefit analysis (Maraniss and Weisskopf 1995b).

The career paths and ideologies of Chao, Scalia, and Strunk reveal the strong policy-planning links between corporations, conservative think tanks, and the highest reaches of the Republican Party. All three were intimately linked to Business Roundtable members American Express, Anheuser-Busch, and UPS, and served to convey the Business Roundtable's position to policy makers that the "government should work with the private sector to . . . encourage autonomy and not impede innovation; and support private sector efforts to improve workplace safety and health" (Business Roundtable 1993, 3).

In addition, corporate-financed think tanks such as the Cato Institute, the Heritage Foundation, and the American Enterprise Institute also played important roles in setting the policy agenda along corporate lines. In the media, on Capitol Hill, through publications and reports, and at public forums, they promoted the view that the WMSD crisis is little more than an employee "comfort" problem that is best dealt with by employers on a voluntary basis.

Budget Battles in Congress

The Democratic Party controlled Congress during Clinton's first two years in office and this gave OSHA a modicum of autonomy from its political opponents, but progress on the standard was slow. The Republican Party captured control of Congress in the 1994 elections for the first time in forty years, and it made the proposed ergonomics standard a top priority for elimination. Working closely with business lobbyists to roll back regulations, House Republicans slated OSHA for a $16 million budget cut and threatened it with further cuts if it did not stop working on the ergonomics standard. In response, the Clinton administration made four major concessions. First, OSHA narrowed the scope of the standard's jurisdiction to apply only where a worker was diagnosed with a WMSD. Up until this point OSHA regulations were designed on a preventive basis. But this did not satisfy House Republican leaders Speaker Newt Gingrich (R-GA) and majority whip Tom DeLay; they ordered another $3.5 million cut from OSHA's already meager budget. According to DeLay, this was equal to the amount that it would have cost business to implement the ergonomics standard, and he added that it was necessary to punish OSHA for "flouting the will of this Congress" (Swoboda 1995). Second, OSHA proposed a "grandfather" clause that would exempt companies with active ergonomics programs, but this concession did little to appease small and midsized firms since most of them did not have ergonomics programs. Third, despite ample evidence of WMSDs among construction, agricultural, and maritime workers, these industries were not covered by the standard. Finally, under the weakened proposal, employers would no longer be required to analyze their safety logs and worker's compensation records in order to identify jobs that might be associated with high repetitive strain injuries (RSI) risks. Nor would they have to evaluate the success of their safety programs to find better ways to prevent WMSDs.

Republicans made effective use of their majority status in Congress from 1995 to 2001 to hold the Labor Department's appropriation bill hostage in order to force OSHA to stop working on the ergonomics standard. In 1995, the House Republican leadership slated OSHA for a 15.5 percent funding cut for fiscal year 1996, but Democrats and moderate Republicans managed to limit the budget cut to only 2 percent. Still this was a blow to OSHA, which is a chronically underfunded agency. Moreover, the Republicans had attached an amendment, called the "ergo rider," to the Labor Department's final appropriations bill. Introduced by DeLay's Texas ally, Representative Henry Bonilla (R), it prohibited OSHA from issuing a standard, or even voluntary guidelines, on ergonomics during the coming fiscal year. It was small consolation to organized labor that it had defeated a more restrictive version of the rider favored by the NCE that would have prohibited OSHA from conducting any work on ergonomics, or even collecting data to assess the extent of the problem. Representative Ballenger summed up House opposition to the ergonomics standard when he cavalierly said: "No one ever died of ergonomics" (Swoboda 1995). Perhaps he was more concerned about the economic impact of the ergonomics standard on the chicken processing companies in his state of North Carolina. Poultry workers, who face work speed-ups under assembly-line conditions, suffer from one of the highest rates of WMSDs.

In July 1996, organized labor won an appropriations battle in the House when Bonilla's ergonomics rider came up for renewal. House Democrats offered an amendment by Representative Nancy Pelosi (D-CA) to strip Bonilla's restrictive ergo rider from the labor appropriations bill that passed by a vote of 216–205. The AFL-CIO's campaign to unseat antilabor Republicans in 1996 persuaded thirty-four moderate House Republicans to distance themselves from Newt Gingrich, who had threatened to shut down the federal government as he had done the year before (Mogensen 2000, 255). But labor's victory was short-lived. In 1997, congressional Republican leaders wooed back moderate GOP and conservative Democratic members with the promise that they would not delay the ergonomics standard for a second time. But the Republicans soon broke their promise, and the restrictive Bonilla rider was attached to the labor appropriations bill again in 1997. Meanwhile, the American Trucking Association increased its lobbying pressure on Capitol Hill by releasing a report claiming that compliance costs could reach $6.5 billion annually for the trucking industry alone. While the cost estimate was wildly overinflated, it was effective in corralling moderate Republicans who might stray from the Republican fold again (American Trucking Association, 1996, 1).

Clinton's Mixed Signals

OSHA occupied a precarious position where it was permitted to resume work on the standard but lacked strong political backing from the White House. The fate of the ergonomics standard depended largely on Clinton's willingness to support it. In April 1996, he addressed the SEIU convention and promised to protect worker safety and health. "We can't afford to jeopardize the future of working Americans by undermining the safety and the solidarity of the workplace, and if such legislation crosses my desk, I'll do what the Constitution entitles me to do: I'll veto it," he said (Clinton 1996, 627). But Clinton had failed to veto the labor appropriations bill containing Bonilla's restrictive ergonomics rider the year before.

Clinton told labor that he would protect workers' safety rights, but he also talked about the need to "reinvent government," which is political jargon for downsizing government and the privatization of public services. Declaring in his 1996 State of the Union address that "the era of big government is over," Clinton campaigned for reelection on the relatively small, inexpensive accomplishments of his first term. He told the SEIU that "our Government is now the smallest it's been since 1965, but it's still strong enough to protect workplace safety" (Clinton 1996, 626). Clinton's comments may have been designed to attract moderate Republican voters, but they raised doubts in the labor community about the strength of his commitment. Clinton's plan was similar to the Reagan and Bush administrations' plans in that it relied on good-faith efforts by employers to report injuries and illnesses while it offered the tempting prize of exemption from inspection to companies that reported fewer incidents. While some companies might follow the honor system voluntarily, historical experience with pressures to maximize profits under market conditions indicates that spending money on ergonomics programs will not be a high priority for most businesses. Indeed, the Insurance Information Institute attributed the stabilization of RSI rates in the meatpacking industry largely to the fact that companies were implementing ergonomics programs based on OSHA's ergonomics guidelines issued in 1991 (King 1996). The meatpacking industry was compelled to comply with OSHA's guidelines by the crisis of rising WMSDs and large OSHA fines, not simply because of voluntarism.

Political Science

Opponents of OSHA's ergonomics standard were guilty of practicing a double standard regarding the science of ergonomics. In public they tried to undermine the need for an ergonomics standard by challenging the legitimacy of the science of ergonomics. Conservative think tanks, corporate lobbyists, and Republican politicians were effective in creating an echo chamber effect by repeating the false charges that ergonomics was "junk science" and that WMSDs did not exist. But privately they accepted and utilized the science of ergonomics when it was to their advantage. Key opponents said that OSHA regulations should be based on sound science but they were unwilling to accept the fact that the National Academy of Sciences (NAS) studies found that the ergonomics standard was based on sound science. In 1993, the Business Roundtable laid the ideological groundwork for a renewed corporate attack on OSHA regulations declaring that "the nation should focus its limited safety and health resources on real, not hypothetical, workplace risks that are identified through sound, state-of-the-art science" (Business

Roundtable 1993, 3). The corporate spin was that OSHA had selectively "cherry-picked" data to support its claim that the ergonomics standard was needed. They even made the specious claim that the science of ergonomics itself was unscientific. Despite decades of documented research and proven results, conservative members of Congress like Senator Fred Thompson (R-TN) repeated the corporate spin that ergonomics is "a field for which there is little if any credible evidence" (Congressional Record 2001, S1845).

The goal of this corporate strategy was that if ergonomics was not a science then WMSDs did not exist. If WMSDs do not exist then they must be the imaginings of hysterical workers. As Edward Shorter, professor of the history of medicine at the University of Toronto said in 2001: "The whole RSI thing has sort of evaporated in a cloud of smoke." He added: "The fact is that most of these people didn't have carpal tunnel syndrome. They had hysteria." But hysteria cannot account for the 1.8 million incidences of WMSDs that are being reported annually in the United States alone. Furthermore, Shorter concluded: "And if you have basically hysteria and you can dignify it with something that makes it sound like you have a serious illness, then you get more respect from people" (Evenson 2001). While Shorter did not mention women by name, they suffer from a higher rate of WMSDs than men. There is a long history of labeling women as hysterical so as not to take their health problems seriously (Dembe 1996). The use of the word "hysteria" in the context of a discussion of WMSDs could be interpreted as a veiled reference that women who reported their injuries were being hysterical. In addition to denying the well-documented existence of WMSDs, the story also confused the terms carpal tunnel syndrome and repetitive strain injuries by using them interchangeably, although the former is but one of over twenty types of the latter: "Today, the [carpal tunnel] syndrome has all but disappeared from Canadian offices," it reported. It is not surprising that this story, which echoed the corporate spin, appeared in *The National Post,* a Canadian newspaper founded by Conrad M. Black, who, like his rival Rupert Murdoch, runs a media empire that reports the news with a distinctly conservative bias.[5]

Despite the public pronouncements of business interests that there was little or no scientific basis to WMSDs, in private they readily utilized the science of ergonomics. The insurance industry, which fought against the ergonomics standard in the public arena and filed a federal lawsuit to stop it, turned around in private and advised its client companies to cut their insurance premium costs by preventing WMSDs through use of the science of ergonomics. It is impressive to note that at least 30 percent of U.S. corporations have voluntary in-house ergonomics programs to prevent WMSDs from occurring, and many send representatives to safety conferences to educate themselves about good ergonomics practices.

Dramatic evidence that corporate America acknowledges that WMSDs are a serious problem and tried to use it to further their interests at the expense of workers' health, came from the Burlington Northern Santa Fe Railroad (BNSF). In a controversial case, it conducted genetic tests on employees who had submitted workers' compensation claims for work-related cases of carpal tunnel syndrome without their knowledge or consent. The purpose of this DNA testing was to determine whether or not these workers had a genetic predisposition to work-related carpal tunnel syndrome. If so, BNSF planned to challenge the disabled workers' claims for compensation on the grounds that their injuries were not work related. Genetic or nonwork-related injuries are not generally covered by workers' compensation. BNSF was forced to stop its genetic testing program when the U.S. Equal Employment Opportunity Commission (EEOC) determined that the railroad had violated the Americans with Disabilities Act by discriminating against their disabled employees. The EEOC's investigation determined that the BNSF had acted "with malice or reckless disregard" for the rights of workers, and that it had punished employees who refused to submit to the genetic tests (Koenig 2001). This incident gave new meaning to BNSF's declaration that its "safety vision is to operate injury- and incident-free" (BNSF 2004, 1). Meanwhile, BNSF, which opposes an ergonomics standard and favors freer trade with Mexico, maintains strong ties with Capitol Hill by sponsoring events for key leaders, such as its lavish luncheon for the wife of House Speaker Dennis Hastert (R-IL), held at the Central Park Boathouse during the 2004 Republican National Convention.

Eugene Scalia, the leading lobbyist in the campaign to defeat the ergonomics standard, was probably the most influential purveyor of the corporate spin that ergonomics was a "junk science." In a 1994 he wrote a white paper for the National Legal Center for the Public Interest titled: "Ergonomics: OSHA's Strange Campaign to Run American Business." A corporate lawyer, not a scientist, he sarcastically dismissed the science of ergonomics, saying: "Like a cruise through Disney World's Pirates of the Caribbean, to survey ergonomists' theories is to glimpse the exotic and absurd, occasionally amusing, and sometimes grisly" (Scalia 1994, 8). According to the corporate logic, if ergonomics is not a science, then there is a "distinct possibility that that there is no such thing as RSIs" (Scalia 1994, 14). He repeated these distortions in a 2000 policy commentary written for the Cato Institute. He alleged that OSHA wants to entrench the questionable science of ergonomics in a permanent rule. But no agency should be permitted to impose on the entire American economy a costly rule premised on a 'science' so mysterious that the agency itself cannot fathom it" (Scalia 2000, 1).

But the real mystery is why Scalia and other opponents of the ergonomics standard overlooked the overwhelming body of evidence that ergonomics is a science. On two separate occasions congressional Republicans asked the prestigious NAS to evaluate whether or not OSHA's ergonomics regulation was based on sound science. On both occasions it answered in the affirmative. The first NAS study, *Work-Related Musculoskeletal Disorders: A Review of the Evidence,* was released in October 1998. It was the work of a panel of sixty-five ergonomics experts from around the world who were invited to Washington, D.C., to consider the scientific evidence. They concluded that poorly designed workplaces caused WMSDs and that ergonomics interventions help rectify these problems (NAS 1998, 1). Other groups weighed in on the evidence behind ergonomics. The National Advisory Committee on Occupational Safety and Health, the body of safety and health specialists from academia, labor, and management that advises the U.S. Secretary of Labor, concluded that "The science of ergonomics is strong and dates back at least fifty years and, in our view, is sufficient to move forward with a proposed OSHA ergonomics standard. Clear evidence of this assertion is that many countries and *Fortune* 500 companies are using ergonomics successfully to reduce workplace injuries" (House 1999).

Actually, Scalia's denial that ergonomics is a science is no mystery at all. It was an integral part of the corporate policy-planning network's organized effort to deny the science behind the ergonomics standard long enough that it could be killed. Both think tanks that published and promoted Scalia's diatribes, the National Legal Center for the Public Interest and the Cato Institute, are part of the corporate policy-planning network. As befits the deregulatory ideology of their directors and advisers who circulate between the power elite of corporate America and the upper echelons of the Reagan and Bush administrations, both institutions share the mission of promoting free enterprise, private property rights, privatization of government services, and limited government. It is a laissez-faire ideology that leaves little room for the scientific process, the regulation of business, the protection of workers' rights, or the public interest.

This stubborn adherence to ideology in the face of scientific proof to the contrary is illustrated by the fact that the first NAS report, which validated the science of ergonomics and OSHA's methodology, did nothing to deter congressional Republicans from warning President Clinton not to release the ergonomics standard until the release of NAS's second report in January 2001. Nor did the NAS report stop them from repeating the unfounded assertion that the ergonomics standard was based on "junk science."

In January 2001 the National Academy of Sciences released the second study commissioned by the congressional Republicans. Once again, it concluded that OSHA's standard rested on a solid foundation of more than 2,000 soundly conducted scientific studies of workplace conditions. Once again, it concluded that there is a substantial body of scientific evidence that links musculoskeletal disorders of the lower back and upper extremities to particular jobs and working conditions—including heavy lifting, repetitive and forceful motions, and stressful work environments (NAS 2001). But by then it was a moot point. The Republicans were not interested in science; they were more interested in "political" science that had served their purpose of delay and denial well.

But if Scalia is correct, how did organized labor, which has less influence in Washington than corporate interests, foist such a charade on the American public, government, and the scientific community? How did it get the Bush administration to allow work on the ergonomics standard to begin if it was all based on fraud? The answer was that the ergonomics standard was a legitimate response to the WMSD problem. The heart of the matter was money, not science. As Scalia himself concluded, the ergonomics standard "means an intrusion on business management and an increase in corporate costs unprecedented in the regulation of American industry" (Scalia 1994, 38).

Despite corporate claims that ergonomics was "junk science," the science behind the ergonomics standard was backed by a broad consensus of the medical and public health community. APHA's Occupational Health and Safety Section sent an urgent letter to Congress during the summer of 1996 criticizing the GOP's attempt to renew the restrictive ergo rider: "Within the scientific community, there is a strong consensus based on an extensive body of solid evidence, on the role of ergonomic factors in the incidence of workplace injuries." They added that: "If no more data are collected, workplace injuries will erroneously appear to be eliminated. Prevention programs will be endangered and important research terminated. Crippling injuries throughout American workplaces will be the inevitable result." APHA warned its allies that "This year, the opponents are better organized, and are circulating a document purporting to show there is no scientific basis for the ergonomics standard. We can stop this legislation in the Senate again, but to do so, it is very important that the scientific and professional occupational health community respond to this new attack" (APHA 1996, 1).

In July 1997 NIOSH rebutted the unfounded charges that ergonomics was junk science by releasing *Musculoskeletal Disorders and Workplace Factors*. A literature review of more than 600 studies, it concluded that "A large body of credible epidemiological research exists that shows a consistent relationship between MSDs and certain physical factors, especially at higher exposure levels" (NIOSH 1997, 1).

Despite the scientific evidence supporting ergonomics, Bonilla's ergo rider was passed for the second time, and again OSHA was prohibited from working on the standard and doing scientific research on ergonomics during the fiscal year 1997. This intense congressional lobbying pressure forced OSHA to make further concessions. First, it postponed the release of its draft ergonomics standard for public comment from June to September 1998. The emergence of "new ideas" that had to be taken into account was OSHA's euphemistic way of explaining effective business opposition. As a result, at the September 1998 stakeholders' meeting OSHA staff seriously considered limiting the standard to manufacturing and manual handling sectors with the "highest reports of injury and known solutions." Labor representatives responded that all work should be covered, saying that it would be confusing and made no sense to include some parts of an industry and not others. A number of participants made a strong case for including computer workers; the Communications Workers of America provided data from studies they conducted showing high WMSD rates among telephone operators. Speaking off the record, OSHA administrator Charles Jeffress said that VDTs would be covered, but as corporate pressure increased, OSHA retreated from the idea of a standard that would cover all industries (Lichterman 1998). With the 1998 congressional elections approaching, the Clinton administration did not want to encourage the wrath of the House Republican leadership, so they postponed the release of the draft standard until early 1999.

When OSHA finally issued its first draft in February 1999, House Republicans moved quickly to try to kill it. In March Representative Roy Blunt (R-MO) introduced a bill to prohibit OSHA from issuing the ergonomics standard, or guidelines, until completion of another two-year study by the NAS. This was the same tactic of delay that worked two years earlier for Bonilla and it worked again. The NAS report did not provide corporate and conservative opponents with the scientific support they sought, but it did give them a precious political commodity—time. In August 1999 the House of Representatives passed the Blount bill, preventing OSHA from issuing the ergonomics standard for eighteen months. Given that the NAS had just completed a similar study requested by Republicans in 1997 verifying that the ergonomics standard was based on scientifically sound data, the only reason for another study was to stall for time until after the 2000 presidential election (NAS 1998). But the GOP's real concern was with the political, not the scientific, process. The Republicans' plan was to delay the standard's release until after the 2000 presidential election in the expectation that a Republican would be in the White House. The Republican front-runner, George W. Bush, had made it clear that, as president, he would make deregulation one of his top priorities. While the Senate did not pass the Blount bill, the process served the purpose of further delaying the standard's release and intimidating OSHA.

The magnitude of the Bush administration's manipulation of the science behind occupational safety and health exceeds that of any previous administration. In February 2004 the Union of Concerned Scientists (UCS) released the results of its investigation into the Bush administration's abuse of scientific procedure to support its political positions on various government programs. Endorsed by 48 Nobel laureates, 62 recipients of the National Medal of Science, 127 members of the National Academy of Sciences, and an additional 4,000 scientists, the report, titled *Scientific Integrity in Policymaking,* concluded that "there is significant evidence that the scope and scale of the manipulation, suppression, and misrepresentation of science by the Bush administration are unprecedented" (UCS 2004, 2). The Union of Concerned Scientists' report also found that "there is a well-established pattern of suppression and distortion of scientific findings by high-ranking Bush administration political appointees across numerous federal agencies. These actions have consequences for human health, public safety and community well-being" (UCS 2004, 2).

The month before the Union of Concerned Scientists released its report, eleven ergonomists boycotted OSHA's National Advisory Committee on Ergonomics symposium charging that the Bush administration was suppressing and manipulating scientific findings that did not agree with its political position. The boycott was a reaction to Secretary of Health and Human Services Tommy Thompson's unprecedented act of rejecting three ergonomics experts who were slated to serve on NIOSH's ergonomics advisory panel, despite the fact that they had been approved by NIOSH for the positions, that they were well qualified for the positions, and that they were highly respected among their peers. Thompson's action violated the spirit of the Federal Advisory Committee Act that requires advisory committees to include balanced representation of different scientific views and prohibits undue interference in the selection process by political appointees or interest groups. A NIOSH staffer said in confidence that he "had never before seen this kind of decision coming in contravention of the agency's recommendation" (UCS 2004, 24). The Union of Concerned Scientists' investigation concluded that Thompson's decision was part of "a wide-ranging effort to manipulate the government's scientific advisory system to prevent the appearance of advice that might run counter to the administration's political agenda" (UCS 2004, 2).

Corporate interests sought to use science to bolster their claims that the causes of WMSDs did not exist or were beyond the employer's control. Behind these highly politicized efforts to deny the existence of ergonomics and WMSDs is the common desire by employers to externalize the costs of doing business by passing them on to labor and society. Economists euphemistically call the socially destructive side effects of production *externalities* because capitalist economics does not consider them to be part of the cost of doing business. They are more accurately called social costs because workers and society must bear the cost.

Corporate Contributors Commoditize Access and Influence

The corporate attack on the ergonomics standard was fueled by massive amounts of money. As with its approach to other public goods, corporate contributions have the effect of commodifying access and influence of public policy. Members of Congress receive many calls for legislative assistance, but their dependence on campaign contributions ensure that they will give higher priority to interests that contribute the most money. During the 2000 election cycle the thirteen leading corporate opponents of the ergonomics standard gave members of Congress $14,879,297, and 83 percent, or $11,306,753, of that went to the Republicans who run Congress. Not one of the corporate interests on the list gave a majority of its contributions to the Democrats, and although the American Bakers Association gave the smallest amount, it gave every penny of its $143,300 to the Republicans. Three of the top contributors to the campaign to defeat the ergonomics standard were shipping and trucking companies and their trade association, which employ warehouse and delivery workers who are at high risk of developing WMSDs. UPS, which was struck by its Teamsters Union–represented workers in 1997 over issues such as load lifting limits and WMSDs, topped the list at $2,918,969, with Federal Express a close second at $2,578,978, and the American Trucking Association fifth at $1,097,563. Sandwiched in between was the brewing industry, which also employs warehouse and delivery workers; the National Beer Wholesalers came in third with $2,124,911, followed by Anheuser-Busch with $1,681,075. The two trade associations that coordinated the lobbying attack on the ergonomics standard were the NFIB, at $1,062,142, and the U.S. Chamber of Commerce, at $516,249. While they did not give the largest amounts, it is illustrative of the intensity of their ideological zeal that they gave 97 percent and 94 percent of their contributions, respectively, to the Republicans (CRP 2000, 1).

The American Insurance Association gave $490,386, 91 percent of it to the Republicans (CRP 2000). The insurance industry objected to the fact that the ergonomics standard would establish a nationwide compensation system for WMSD victims alongside the established framework of state workers' compensation systems. The insurance industry dominated the state systems and saw the new system as a potential threat to the status quo. Its elected allies argued that establishing a national WMSD compensation system was a violation of states' rights, but the bottom line was that the national system mandated higher payments than state laws provided, which could cut into the insurance industry's profit margins.[6]

Promulgation and Repeal

While some major corporations were willing to accept the watered-down ergonomics standard if necessary, competitive capital would not compromise. House Republicans learned this lesson the hard way on the eve of the 2000 presidential election. House GOP conservatives had threatened to shut down the federal government if Clinton released the standard before the election. But the Congressional Republican leadership did not want to be blamed by angry voters on the eve of the presidential election. The example of Newt Gingrich being blamed by public opinion for shutting down the federal government a few years before must have been fresh in their memory. As a result, they reached a tentative behind-closed-doors compromise with the Clinton administration to let the standard go into effect in June 2001. The House Republican leadership was willing to gamble that George W. Bush would be the next president, and if so, with Congress's cooperation, he could repeal it, but business interests feared that the compromise would tie the incoming president's hands. The tentative deal fell apart the next day when its details were leaked to the media and the U.S. Chamber of Commerce and the NCE vehemently objected to it (Cross 2000, 1). The NCE said: "The proposed ergonomics 'compromise' represented no compromise at all on the part of the Administration. As reported, it would have allowed the issuance of a final ergonomics regulation, without regard to the fact that a new Administration might wish to rescind or otherwise modify the rule prior to its final promulgation" (NCE 2000, 1).

The next move in this legislative cat and mouse game occurred when Congress recessed for the 2000 presidential election. With the members of Congress back in their districts, the Clinton administration pounced on the opportunity and promulgated the final standard on November 14, 2000, by publishing it in the *Federal Register.* Although it was ten years in the planning, the ergonomics standard was short-lived. Bush's first official act as president, taken on the same day he was inaugurated, was to sign an executive order freezing work on all pending regulations for two months. The executive order was written by his chief of staff, Andrew H. Card, Jr., a Bush family friend and member of the policy-planning network. His career provides yet another example of the revolving door between the commanding heights of corporate America and the upper echelons of the federal government. Prior to becoming Bush's chief of staff, Card was General Motors' vice president of governmental relations. He served as the elder President

Bush's transportation secretary from 1992 to 1993, and directed the administrative transition to the Clinton administration. Earlier, Card served in the Bush White House as assistant to the president and deputy chief of staff. During the Reagan administration, he served in several posts, including director of intergovernmental affairs; and, Card was president and chief executive officer of the American Automobile Manufacturers' Association from 1993 to 1998.

Card's executive order stopped work on all pending regulations, but another political tactic was necessary to kill the ergonomics standard that had gone into effect four days before Bush took office. Republican leaders in Congress found their answer in an obscure law called the Congressional Review Act (CRA). Passed by a Republican-controlled Congress and signed into law by President Clinton in 1996, the CRA gives Congress and the president the unprecedented power to overturn newly enacted regulations. Under the CRA's stringent provisions, the agency that issued the regulation is prohibited from promulgating it again without congressional authorization. In effect, it would be virtually impossible for OSHA to reissue an ergonomics standard of commensurate strength in the prevailing antiregulatory political climate. The AFL-CIO description of the CRA as the "neutron bomb of deregulation" was apt. On March 6, 2001, the Senate voted 56–44 to repeal the ergonomics standard and the House of Representatives did the same the next day, voting 223–206. It marked the first and only time that the Congress has used the CRA to pass "resolutions of disapproval" (Congressional Record 2001). On March 20, President Bush fulfilled a campaign pledge to his corporate supporters by signing the ergonomics standard's repeal into law.

Ironically, President Clinton sowed the seeds that undid his greatest labor achievement. Why did Clinton sign the CRA into law? The answer is twofold. First, it was Clinton's willingness to move to the right and co-opt Republican-dominated issues, such as balancing the budget, cutting welfare benefits, and "reinventing government" by cutting the size of the federal bureaucracy. Second, the congressional Republicans had shut down funding for the federal government and passage of the CRA was one of Gingrich's demands for passing the budget resolutions. Clinton may have acquiesced to Gingrich's demand thinking that the chances of such a draconian legislative instrument being used were extremely remote. Since both Congress and the president must agree to repeal a regulation, its likelihood of being used is greatly diminished when divided government prevails, or the proregulatory Democrats control both ends of Pennsylvania Avenue. For the CRA to be used effectively, antiregulatory forces need to control both Congress and the presidency, and that situation occurred in an improbable way when George W. Bush was effectively installed in the White House by the five staunchest

Republican appointees on the U.S. Supreme Court. However, the fact remains that Clinton traded the lasting effects of CRA for short-term advantage in 1996, and he must bear the responsibility for handing regulatory opponents such a destructive weapon.

Ergonomics After the Fall

To gain support for the ergonomics standard's repeal from undecided members of Congress in March 2001, Secretary of Labor Chao promised that the Bush administration would "pursue a comprehensive approach" to ergonomics that may include new rulemaking (Sutcliffe 2001). However, two months later she would not commit to a two-year rulemaking deadline sought by some senators (Wislock 2001). The business community, thankful for the standard's repeal, did not object. Ed Gilroy, head of the NCE, and the American Trucking Association praised Chao at Senate hearings in July 2001: "The process laid out by the Secretary is both fair and even-handed. And, the questions posed are objective and reasonable" (NCE 2001, 1). But he had nothing but scorn for organized labor that protested the ergonomics standard's repeal: "Unfortunately, the AFL-CIO seems more intent on political grandstanding, rather than participating in an honest discussion of the fundamental issues that persist with respect to ergonomics" (NCE 2001, 1).

Karen Bosgraaf, the NFIB's manager of regulatory policy said: "The Bush administration is going to focus more on compliance assistance, on a partnership with small business." On the Bush's administration's promised rule, she added: "Hopefully they'll be less heavy handed and costly, and focus more on voluntary efforts" (Harper 2001). Bosgraaf's comments proved to be prophetic; the Bush administration favored the voluntarist approach. Bush announced a new committee stacked with business friendly appointees to advise the Department of Labor on WMSD research, a plan that bypasses NIOSH's statutory role as the research agency for OSHA. Not surprisingly, when the National Advisory Committee on Ergonomics finally met in January 2003 there was no mention of enforcement of WMSD violations and the Bush administration made it clear that the regulatory approach was off the table. John Henshaw, Bush's OSHA administrator, told the committee that "enforcement, per se, will not be part of your deliberations." To stifle any discussion of reviving the regulatory process, he made it clear that, despite the Bush administration's promises, an ergonomics standard "should not be a part of the committee's discussions" (Nash 2003, 1). The Bush administration still has not announced a proposal for a new standard. Instead, it continues to rely on voluntary guidelines, which lets the employer decide what it will or will not do.

Taking its cue from Scalia and the corporate agenda, the Bush administration acted as if WMSDs were not an occupational safety and health problem. It refused to use the general duty clause until 2003 to cite worst-case violators; even the Reagan and George H.W. Bush administrations were more willing to use the general duty clause in similar circumstances. But even the Bush administration could not deny the existence of WMSDs forever. Ironically, its first use of the general duty clause was against the small- to medium-size business sectors, the very interests who were most vociferous in their claim that WMSDs were not a problem. In March 2003 OSHA cited Alpha Health Services, Inc., an Idaho-based nursing chain, and Security Metal Products Corporation, a metal fabricating company, for ergonomics violations that had occurred. They paid reduced fines of $2,200 and $51,300, respectively (US OSHA 2003, 1; "It's Enforcement" 2003, 12–13).

Further embarrassing evidence that the ergonomics problem would not go away came from a corporate ally of the Bush administration. In 2002 Liberty Mutual released its annual survey of the causes of occupational injuries and found that two ergonomics-related problems made its top-ten list: overexertion ranked first and repetitive motions ranked sixth. Senator John Breaux (D-LA) introduced a bill in 2002 that would give OSHA two years to issue a new ergonomics standard. His version was similar to the repealed standard minus the national workers' compensation proposal; instead, the states would be encouraged to expand their plans. However, Breaux's bill was effectively killed in the Senate.

The revolving door between the Bush administration and the business community, via the policy-planning network, continues to operate to frustrate efforts to improve job safety conditions. In October 2003, President Bush used the recess appointment option to nominate Gary Lee Visscher to be a member of the U.S. Chemical Safety and Hazard Investigation Board. Democrats and organized labor opposed Visscher's nomination on the grounds that he, a corporate lawyer and lobbyist, did not meet the statute's qualifications for technical expertise. Like Scalia's case, the fact that Bush had to resort to appointing Visscher without a hearing—although the Republicans controlled the upper chamber—indicates the level of disregard for his ability to perform the job impartially. But Visscher proved his bona fides to the White House through his opposition to the ergonomics standard. Like other Bush appointees, Visscher's career illustrates the close links between the administration and corporate interests. A deputy assistant secretary for Occupational Safety and Health at OSHA at the time of his appointment to the Chemical Safety and Hazard Investigation Board, he was vice president of Employee Relations for the American Iron and Steel Institute, a staunch opponent of OSHA and the ergonomics standard.

Prior to that, he was appointed to serve as a member of the Occupational Safety and Health Review Commission.

Yet another corporate lobbyist who helped to kill the ergonomics standard was rewarded with a position in the Bush administration's Labor Department. Ed Frank, appointed to be assistant secretary for labor, was the point man in the effort to tighten the rules to make it more difficult for workers to qualify for overtime pay, and as a result, an estimated 6 million workers no longer qualified for overtime pay. Frank previously worked as the NFIB's spokesman—the interest group that led the lobbying campaign to repeal the ergonomics standard.

Conclusion

The rise and repeal of OSHA's ergonomics standard is best explained by socioeconomic theories of power. The statist approach, in downplaying the importance of socioeconomic factors, is too limited in scope to explain the political dynamics involved in the making and unmaking of the ergonomics regulation. According to the statist approach the prospect of enhancing and expanding the agency's administrative capacity and autonomy provides motivation for OSHA officials to promulgate a new standard. But it was not the decisive factor in the case of the ergonomics standard. Statist theory also argues that only strong state agencies have the capacity to promote their own policies. But how was OSHA, a weak state agency, able to promulgate the ergonomics standard after a ten-year struggle? The answer is to be found in socioeconomic explanations; the ergonomics standard was organized labor's idea and it would not have survived the ten-year struggle with corporate interests without labor's lobbying support. Structuralists argue that the relative autonomy of the state creates the potential for public policies that contradict capital's interests. These policy disjunctures provide opportunities for organized labor to politically exploit, as it did with the ergonomics standard. But capital subsequently succeeded in closing the policy disjuncture with the standard's repeal.

Both the structuralist and instrumentalist approaches shed light on how capital mobilized its resources to close the policy disjuncture caused by the ergonomics standard. The structuralist approach explains the leading role played by small- to midsized business interests. They have long had considerable leverage in the House of Representatives, whose numerous districts and short terms are designed to represent local and regional interests to a greater extent than the Senate. The instrumentalist approach explains how corporate interests mobilized their superior resources through the policy-planning network. As this case shows, influential members of the business elite circulated between the corporate world and policy-making positions in the federal government that enabled them to close policy disjunctures relating to ergonomics policy making.

John J. DiIulio, Jr., Bush's director of faith-based initiatives, resigned after he came to the conclusion that the White House was being run by the political operatives like Karl Rove at the expense of sound public policy making. Usually, the political operatives, whose job is to help get a candidate elected, recede in importance after the president takes office, but not in the Bush White House. DiIulio said, "There is no precedent in any modern White House for what is going on in this one: a complete lack of policy apparatus. What you've got is everything—and I mean everything—being run by the political arm. It's the reign of the Mayberry Machiavellis" (Suskind 2004, 170). DiIulio's comment sheds an insider's light on why the Bush administration was so insistent on applying political litmus tests to government appointees—even to nonpartisan scientific advisory boards.

The ability of corporate interests to mobilize superior resources and utilize policy-planning networks and structural links to the U.S. House of Representatives, gave them the political balance of advantage vis-à-vis organized labor, and enabled them to sustain a long-term, coordinated campaign of opposition. Although small and large businesses disagreed on the value of ergonomics programs, both sectors were opposed to government regulation. The campaign by corporate interests, their political allies, and state officials against the ergonomics standard was part of a larger corporate effort to "reform" OSHA, along with other regulatory agencies, in order to reduce production costs and bring the U.S. economy in line with the global free-market economy. In order to achieve this goal, the labor movement was marginalized, the Democratic Party's historic New Deal coalition was weakened, and the labor force was disciplined by limiting workers' rights.

The twenty-year political struggle between capital and labor over OSHA's proposed ergonomics standard reveals the extent to which the state is a contested terrain, but one that is ultimately dominated by business interests. Divided government, along with organized labor's support, gave OSHA enough autonomy to keep the proposed ergonomics regulation alive during the Clinton administration. It took the strong support of the AFL-CIO to sustain OSHA and the standard through the many years of coordinated attacks by Republican foes and corporate interests. Still, it took the intervention of a lame-duck president to implement the ergonomics standard. The ergonomics standard was Clinton's final payment on the outstanding political debt he owed to organized labor for its strong electoral support.

While structural barriers impede the ability of business interests to directly translate their economic power into political power, they still have the balance of advantage against their adversaries, what Charles Lindblom calls "the privileged position of business" (Lindblom 1977, 170). Corporate interests and laissez-faire ideology so dominate the policy debate on safety and health that even if the standard had not been repealed, they succeeded in watering it down and limiting its jurisdiction. Even if Gore were in the White House, Republicans in Congress could cut OSHA's budget. In the final analysis, corporate interests were able to utilize the structure of state power relations to delay the promulgation of the ergonomics standard for years, and ultimately defeat it. Consequently, workers and the public must bear the social costs of this defeat in the form of injuries and illnesses that could have been prevented.

Notes

1. In medical terms, WMSDs are also called cumulative trauma disorders because the condition progressively worsens with each repetitive motion made in an awkward position. This is why variety in work routines and periodic work breaks are an important part of any preventive strategy.

2. The literature is vast, but some key examples are: Stephen Krasner, *Defending the National Interest* (Princeton, NJ: Princeton University Press, 1978); Theda Skocpol, "Political Response to Capitalist Crisis: Neomarxist Theories of the State and the Case of the New Deal," *Politics and Society* 10 (1980): 155–201; Eric Nordlinger, *On the Autonomy of the Democratic State* (Cambridge, UK: Cambridge, MA: Harvard University Press, 1981); Stephen Skowronek, *Building a New American State: The Expansion of National Administrative Capacities, 1877–1920* (Cambridge University Press, 1982); Peter B. Evans, Dietrich Rueschemeyer, and Theda Skocpol (eds.), *Bringing the State Back In* (Cambridge, UK: Cambridge University Press, 1985); Margaret Weir, Ann Shols Orloff, and Theda Skocpol (eds.), *The Politics of Social Policy in the United States* (Princeton, NJ: Princeton University Press, 1988); Theda Skocpol, *Protecting Soldiers and Mothers: The Political Origins of Social Policy in the United States* (Cambridge, MA: Harvard University Press, 1992).

3. Statists revised their theory about strong states being linked with state capacity in light of the debate with neo Marxists over Social Security. By statist standards, the newly created (in 1935) Social Security Administration had neither a bureaucratic history, nor administrative capacity, nor professional expertise to develop into a strong state organization. They argued that Social Security required "nothing more than collecting taxes from employers, keeping updated individual records, and putting a check in the mail" (Weir, Orloff, and Skocpol 1988, 432).

4. They include Abbot Laboratories, Aetna Life and Casualty, Alcoa, American Express, American International Group, Bank of America, Bechtel, Boeing, Boise Cascade, BP Amoco, Chevron, Dupont, Emerson Electric Co., Exxon Mobil, Ford Motor Co., General Motors, Hewlett Packard, IBM, Kerr McGee, 3M, Motorola, Philip Morris, Quaker Oats, Raytheon, RJR Nabisco, Rockwell, Shell Oil, TRW, Union Carbide, Union Pacific, United Parcel Services, Waste Management Inc., Westinghouse, Weyerhaeuser, and Xerox.

5. Hollinger International, Inc. accused Lord Black and an associate of running a "corporate kleptocracy" and sued them for $400 million that they allegedly embezzled from the company (Fabrikant 2004).

6. The American Insurance Association filed a lawsuit in January 2001 challenging OSHA's statutory authority to require that workers be compensated for their injuries, but it became a moot point when the standard was repealed. The suit was filed by twelve of the largest property and casualty groups representing 152 insurance companies. They were ACE USA, American International Group, Inc., California Association of Non-profits Insurance Services, Chubb, CGel Insurance Group, Fireman's Fund, The Hartford, Kemper, Liberty Mutual, PMA, Royal & Sun Alliance, and The Travelers.

References

AFL-CIO. 1997. *Stop the Pain! Repetitive Strain Injuries: An AFL-CIO Background Report*. Washington, DC: AFL-CIO.

———. 2004. *Chronology of OSHA's Ergonomics Standard and the Business Campaign Against It*. Washington, DC: AFL-CIO.

American Public Health Association (APHA). 1996. David Michaels, Chair, Occupational Health and Safety Section, "Letter to Congress on the Scientific Basis for an Ergonomic Standard," June 24.

American Trucking Association. 1996. *Ergonomics and Economics: The Impact of OSHA's Proposed Ergonomic Standard on the U.S. Trucking Industry.* Prepared for ATA by National Economic Research Associates, October.

Burlington Northern Santa Fe Railroad (BNSF). 2004. *Employee Safety Programs*. Available at: www.bnsf.com/about_bnsf/employeesafety04.pdf.

Business Roundtable. 1993. *Occupational Safety and Health: General Principles*. Washington, DC: Business Roundtable, 3.

Center for Responsive Politics (CRP). 2000. *Big Business Influence on Congressional Ergonomics Opponents: Election Cycle 2000 Political Contributions*. Washington, DC: Center for Responsive Politics.

Clinton, William J. 1996. *Public Papers of the Presidents*, Volume 1. Washington, DC: General Printing Office.

Croasmun, Jeanie. 2004. "Women Much More Likely to Develop MSDs." Ergoweb. July 26. Available at: www.ergoweb.com.

Cross, Al. 2000. "Clinton: Voters Have 'Clear Choice.'" *Louisville Courier-Journal* (November): 1.

Dembe, Allard E. 1996. *Occupation and Disease: How Social Factors Affect the Conception of Work-Related Disorders*. New Haven, CT: Yale University Press.

Domhoff, G. William. 1996. *State Autonomy or Class Dominance? Case Studies in Policy Making in America*. New York: Aldine de Gruyter.

Evenson, Brad. 2001. "Repetitive Stress Pain Was Just 'Hysteria': Fad Diseases Cause Real Pain, But Not the Way We've Been Told, Researcher Says." *National Post* (Canada), June 9.

Fabricant, Geraldine. 2004. "Hollinger Files Stinging Report on Ex-Officials." *New York Times*, September 1.

Harper, Philipp. 2001. "Workplace Rules Numbed, But Business Will Feel Other Pains. Has a Derailment Ever Been So Welcome?" Available at: www.msnbc.com.

"It's Enforcement, But Not as We Know It." 2003. *Hazards* (April–June): 12–13.

Jeffress, Charles N. 2000. Statement of Charles N. Jeffress, Assistant Secretary for Occupational Safety and Health, U.S. Department of Labor, Before the Subcommittee on Regulatory Reform and Paperwork Reduction, Committee on Small Business. April 13. Available at: www.osha.gov/pls/oshaweb/owa-disp.show_document?p-table:\TESTIMONIES&p.id=224.

"Job Redesign Needed But Must be Done by Employers Themselves." 1986. *Occupational Safety and Health Reporter* (OSHR). May 29.

Kilborn, Peter. 1989. "Rise in Worker Injuries is Laid to the Computer." *New York Times*, November 16.

King, Mason. 1996. "OSHA Still Considering Repetitive-Stress Rules; Opponents Say Guidelines Aren't Needed." *Indianapolis Business Journal*, July 22.

Koenig, David. 2001. "EEOC: Burlington Testing Illegal" *Washington Post*, July 20.

Lichterman, Joan. 1998. "OSHA'S Last Ergonomics Stakeholders' Meetings and Next Steps." *The RSI Network*, issue 32, October. Available at: www.ctdrn.org/rsinet/archive/rsinet32-oct98.html.

Lindblom, Charles E. 1979. *Politics and Markets*. New York: Basic Books.

Maraniss, David and Michael Weisskopf. 1995a. "Forging an Alliance for Deregulation; Rep. DeLay Makes Companies Full Partners in the Movement." *Washington Post,* March 12.

———. 1995b. "OSHA's Enemies Find Themselves in High Places." *Washington Post*, July 24.

Miliband, Ralph. 1969. *The State in Capitalist Society*. New York: Basic Books.

Mogensen, Vernon L. 1996. *Office Politics: Computers, Labor, and the Fight for Safety and Health.* New Brunswick, NJ: Rutgers University Press.

———. 2000. "Ergonomic Inaction: Congress Puts OSHA's Ergonomic Standard on Hold." In *Computing for Non-specialists*, ed. Nanda Bandyo-padhyay, 251–55. Harlow, UK: Addison-Wesley.

———. 2001. "RSIs Aren't Real and Other Tales of Voodoo Science." *Hazards*, 75: 4–5.

Nash, James L. 2003. "Ergonomics Advisory Committee Meets—Enforcement is Off the Table." *Occupational Hazards*, January 24. Available at: www.occupationalhazards.com.

National Academy of Sciences (NAS) 1998. "National Research Council. Steering Committee for the Workshop on Work-Related Musculoskeletal Injuries." In *Work-Related Musculoskeletal Disorders: A Review of the Evidence*. Washington, DC: National Academy Press.

———. 1999. *Work-Related Musculoskeletal Disorders: Report, Workshop Summary, and Workshop Papers*. Washington, DC: National Academy Press.

———. 2001. "Panel on Musculoskeletal Disorders and the Workplace. Commission on Behavioral and Social Sciences and Education." In *Musculoskeletal Disorders and the Workplace: Low Back and Upper Extremities*. Washington, DC: National Academy Press.

National Coalition on Ergonomics. 2000. "NCE Applauds Congressional Leaders for Holding Firm on Ergonomics Regulation." News release, October 31. Available at: www.ncergo.org/pressreleasence.htm.

———. 2001. "NCE Calls on Congress, Others to Allow DOL Ergonomics Review to Proceed Unimpeded." News release, July 18. Available at: www.ncergo.org/pr71801.htm.

Poulantzas, Nicos. 1978. *Political Power and Social Classes*. London: Verso.

Scalia, Eugene. 1994. "Ergonomics: OSHA's Strange Campaign to Run American Business." *The National Legal Center for the Public Interest White Paper*, volume 6, number 3, August.

———. 2000. "OSHA's Ergonomics Litigation Record: Three Strikes and It's Out." *Cato Institute Policy Analysis* No. 370, May 15.

Skocpol, Theda, 1992. *Protecting Soldiers and Mothers: The Political Origins of Social Policy in the United States*. Cambridge, MA: Harvard University Press.

Skowronek, Stephen. 1982. *Building a New American State: The Expansion of National Administrative Capacities, 1877–1920.* Cambridge, UK: Cambridge University Press.

Suskind, Ron. 2004. *The Price of Loyalty: George W. Bush, the White House, and the Education of Paul O'Neill.* New York: Simon & Schuster.

Sutcliffe, Virginia. 2001. "Bush Repeals Ergonomics Rules." *Occupational Hazards*, March 21. Available at: www.occupationalhazards.com/articles/2944.

Sweeney, John J., and Karen Nussbaum. 1989. *Solutions for the New Workforce: Policies for a New Social Contract*. Cabin John, MD: Seven Locks Press.

Swoboda, Frank. 1995. "OSHA to Defy House Ban with New Workplace Rules." *Washington Post*, March 20.

Union of Concerned Scientists (UCS). 2004. Scientific Integrity in Policymaking: An Investigation into the Bush Administration's Misuse of Science. Cambridge, MA: Union of Concerned Scientists.

U.S. Congress, House of Representatives. 1989. Committee on Government Operations. Dramatic Rise in Repetitive Motion Injuries and OSHA's Response: Hearing before the Employment and Housing Subcommittee, 101st Cong., 1st sess., June 6.

———. Office of Technology Assessment (OTA). 1995. *Gauging Control Technology and Regulatory Impacts in Occupational Safety and Health—An Appraisal of OSHA's Analytical Approach.* OTA-ENV-635. Washington, DC: U.S. General Government Printing Office.

———. Senate. 2001. *Congressional Record*, vol. 147, no. 28, 107th Cong., 1st sess., March 6, S1848.

U.S. Department of Labor. Bureau of Labor Statistics (BLS). 1994. *Occupational Injuries and Illnesses in the United States by Industry*. Washington, DC: Government Printing Office.

———. Occupational Safety and Health Administration (US OSHA). 2003. "Oklahoma, Idaho Companies Agree to Work with OSHA to Abate Ergonomic Hazards." Trade news release, March 27.

U.S. National Institute for Occupational Safety and Health (NIOSH). 1995. *Cumulative Trauma Disorders in the Workplace: Bibliography*. Cincinnati: Department of Health and Human Services.

———. 1997. *Musculoskeletal Disorders and Workplace Factors*. Cincinnati: Department of Health and Human Services.

———. 2000. *Working Women Face High Risks From Work Stress, Musculoskeletal Injuries, Other Disorders, NIOSH Finds*. Cincinnati, OH: NIOSH.

Weir, Margaret, Ann Shols Orloff, and Theda Skocpol, eds. 1988. *The Politics of Social Policy in the United States.* (Princeton, NJ: Princeton University Press, 1988.)

Wislocki, John. 2001. "Chao: No New Ergo Rule Likely Within Two Years." *Transport Topics*, May 8. Available at: www.++news.com/members/topnews/0007230.html.

III

The Impact of Neoliberalism on Workers' Safety Rights Abroad

Selected Case Studies

8

The 10 Percenters

Gender, Nationality, and Occupational Health in Canada

Penney Kome

Ten percent of adult Canadians have repetitive strain injuries (RSI) severe enough to interfere with their daily lives. This figure is based on a national survey conducted in 2000–01. As Statistics Canada (StatsCan) reported in August 2003, "This marked an increase in the prevalence of RSIs during the late 1990s. In 1996–1997, 8 percent of adults reported the problem, according to the National Population Health Survey. The proportion hit 10 percent in 2000–2001. Work-related activities were most often the cause" (Statistics Canada 2003).

StatsCan also noted that there was a gender difference along with the increase. In 2000–01, men and women were almost equally likely to report an RSI "although since 1996–1997, the percentage of women sustaining such injuries rose faster than the percentage of men. For women, the increase was from 7.9 percent to 10.3 percent, compared with an increase from 8.2 percent to 9.9 percent for men."

Readers with keen memories might notice that Statistics Canada published this report at about the same time that the U.S. Occupational Safety and Health Administration (OSHA) announced it was giving up any attempt to require U.S. employers to report musculoskeletal disorders (MSDs) in their workplaces (Occupational Safety and Health Administration 2003). The U.S. Bureau of Labor Statistics (BLS) still reports on "Strains, Sprains and Tears," and displays a graph with a steep downward slope from more than 1 million reported cases in 1992, to about 670,000 cases in 2001—quite the reverse of the increase that StatsCan found (BLS 2004).

National RSI statistics have never been easy to obtain in Canada where there is no federal agency comparable to OSHA or the BLS. However, StatsCan used its own initiative to add questions about RSIs to its 1996 National Population Health Survey and then subsequently in its 2001 Community Health Survey. In the United States, Fedstats does not show anything like either of these general health surveys available for comparison (U.S. Fedstats 2004).

StatsCan did report on occupational injuries and illnesses for decades, but that duty was phased out of its mandate in the mid-1990s. Similarly, Human Resources Development Canada (HRDC) used to publish national occupational injury and fatality information, with its last report covering the period from 1976 to 1996.

In 1999 the government outsourced data collection to the nonprofit Association of Workers' Compensation Boards of Canada (AWCBC). The AWCBC's first task was to harmonize workers' compensation boards' reporting definitions across provincial jurisdictions. Even so, such statistics as the AWCBC makes available to the public without fee carry a caution that "Variances arise because the acts and regulations administered by jurisdictions are not identical, and because each jurisdiction has unique operating procedures" (AWCBC 2004).

AWCBC's statistics have other problems too. Only its members (the thirteen workers' compensation boards across Canada) can receive most of the data free of charge. Nonmembers, such as the general public, must pay for its publications. For the first few years, even the number of annual occupational fatalities was proprietary information. Second, and more important, the statistics represent only injuries for which workers' compensation boards (WCBs) have approved lost-time claims, thus omitting workers whose claims were refused, or who took sick time to heal, or who simply changed jobs. No one knows how many injuries go unreported.

Even as the AWCBC set out to conglomerate national data from individual WCBs, the WCBs themselves were changing format and mandate. As is the case in many states, Canadian provinces often amend, revise, and revamp their workers' compensation acts with every change of government. Under the New Democratic Party in the early1990s, the British Columbia government fired its entire WCB and rewrote the legislation. When the Liberals came into power in 2001, they rewrote the act again.

In Ontario, the Liberal government was in power from 1985 to 1990. The Liberals supported an innovative tripartite musculoskeletal injury prevention project (MIPP), cosponsored by government, labor, and the business community, which trained more than 32,000 workers to spot hazards in their workplaces and correct them using ergonomics techniques. MIPP produced detailed manuals tailored to specific industries, from manufacturing to retail.

When the Conservatives won the 1995 election, however, they cancelled government backing for the MIPP project. It was subsequently revived, and carries on as a six-hour course at the Workers' Health and Safety Centre, run by the Ontario Federation of Labour. Moreover, the Conservatives turned the WCB inside out and renamed it the Workplace Safety Insurance Board (WSIB)—thus removing all mention of both workers and compensation. Finally, the new Ontario WSIB announced that repetitive strain injuries would no longer be eligible for compensation, but had to retreat in the face of the resulting outcry (WSIB 2004).

Under Premier Mike Harris, the Conservative government slashed budgets for agencies that protect workers and made sweeping changes to programs. In *Consulted to Death,* Winnipeg author and journalist Doug Smith wrote:

> In its first term the Harris government
> Dismantled the Workplace Health and Safety Agency;
> Reduced training requirements for health and safety representatives;
> Cut the budgets of the Occupational Health Clinics;
> Laid off occupational health nurses and physicians;
> Closed the Occupational Health Laboratory and Library;
> Disbanded the Occupational Diseases Panel; and
> Disbanded the joint labour/employer toxic substance standard setting process. (2000, 112)

Bear these background factors in mind when weighing conflicting data. StatsCan's findings that RSIs have increased would seem to be at odds with AWCBC published data, which show that accepted "time-lost" claims have decreased steadily from about 456,000 in 1992 to 359,000 in 2002. (AWCBC 2004) Although more people in the general population say that they are living with injuries sustained at their jobs, fewer are successfully claiming workers' compensation. Perhaps this difference is the best possible argument for continuing and expanding the practice of conducting general health surveys.

StatsCan found that nearly two-thirds (64 percent) of reported RSIs were in the upper body: 48 percent were in the neck, shoulder, wrist, or hand, with another 16 percent in the elbow or lower arm. Only 19 percent were in the back, and 17 percent involved a lower extremity or unspecified body part. "Men were more likely than women to have hurt their arm, leg or back,"

StatsCan reported. "In contrast, a higher percentage of women than men reported injuries to their neck, shoulder or hand. These differences are likely attributable to the types of activities each sex undertakes." When accurate statistics are available, the gender dimension of strain injuries leaps to the eye. And that dimension is constant across borders. StatsCan's findings corroborate what the AFL-CIO reports on its Web site:

> Women make up approximately 44 percent of the workforce, and 33 percent of those injured at work, yet they account for 64 percent of repetitive motion injuries that result in lost worktime (43,671 out of 68,323).
>
> Women experience 68 percent of the carpal tunnel syndrome injuries that result in lost worktime (18,740 out of 27,697).
>
> Sixty-two percent of the lost-worktime tendonitis injuries occur among women workers (8,961 out of 14,445).
>
> In 2000, nearly 155,000 women workers suffered an injury due to overexertion that resulted in time away from work.
>
> Sprains/strains, carpal tunnel syndrome and tendonitis together account for more than half of all lost worktime injuries and illnesses among women. (AFL-CIO n.d.)

Similar statistics prevail in Europe. A Polish study found that women were almost half again as likely to develop strain injuries as men: 3.3 per 100,000, compared to 2.3 per 100,000 for men ("Women's Health n.d.). In 1994, the French Ministry of Labor reported that "globally, female workers are the most exposed to constrained movement, and especially the unskilled in industry (75 per cent of the exposed group)."

Hazards magazine expounded on the study's conclusions: "66 percent of unskilled female workers in the food and agro industries and 58 percent in the clothing industry were found to repeat the same movements at high speed. The retail sector was high risk too: 45 percent of cashiers scan around 20 objects per minute and handle potentially damaging individual and cumulative loads, with 20 percent of items as heavy as 1–7 kg ("Women's Health n.d.).

In 1999, the European Trade Union Congress published an anthology called *Integrating Gender in Ergonomic Analysis* in which editor and McGill University professor Karen Messing argued that:

> Scientists, employers, decision-makers and even women themselves seem to have difficulty in coming to grips with women's work-related health problems. In part, this stems from traditional perceptions of women's work. The widespread belief that women's jobs are safe compared to men's, means that women's health problems are dismissed as an inability to do the job, or 'all in the mind.' This has held back efforts to improve their occupational

health. Prolonged standing which lead to circulatory disorders, or repetitive movements which cause micro-strains, seem much less dangerous than the risk of falling from scaffolding or saw-cut injuries." (1999)

Yet RSIs can be profoundly disabling and disruptive to a worker's life, at home as well as on the job. Workers say that they cannot lift their children, or help them build projects, or play catch with them. Injuries such as carpal tunnel syndrome or tendonitis may make it impossible for parents to help their children with such basic activities as spreading peanut butter on toast or using a crayon to color in a drawing book. A worker with severe tendonitis may be unable to feed himself with knife and fork, or brush her own teeth, or execute a recognizable signature on a check.

These injuries are particularly prevalent and particularly traumatic for women. Indeed, one reason that women are more vulnerable may be that most married women still come home to a so-called "double shift," performing an average of twenty to twenty-five hours of housework a week in addition to their paid work, according to recent studies (Ramos 2003; Juster, Ono, and Stafford 2002).

Here is another point of confusion about strain injuries. Currently, as well as historically, some experts have argued that women's susceptibility is inherent to being female, rather than being a function of the kinds of work that women do. As Karen Messing wrote in *One-Eyed Science:*

> The fact that aging women tend to have more musculoskeletal problems has not escaped the notice of those who look at menopausal women with suspicion. At the 1994 meeting of the International Ergonomics Association, a doctor harangued a plenary session for several minutes, asserting that it had been scientifically proven that all (sic) carpal tunnel symptoms among older women were due to menopause. At a meeting of union health and safety representatives, one described personal conditions which could be spoken of as causes of MSD and used to block compensation: arthroses (actually sometimes due to work), age and menopause. 'A woman developed a tenosynovitis and we lost at the appeal board. [They said] she hadn't done enough repetitive movements, that wasn't the cause and the company doctor came to tell us that it was because she was close to menopause and it was due to that. She packs 25,000 boxes a day.' (a cycle time of less than 2 seconds).
>
> However, the problems of younger women can also be attributed to hormones. If you're a young woman and you were pregnant during the year or the year before and you have problems with carpal tunnel they will often say it's because you have had pregnancies. And they will often bring in a doctor who will say that a woman who has had a child this year

> or the year before, it's common, it's normal that would be the cause of her carpal tunnel.
>
> According to workers, menopause and pregnancy are often brought up in compensation hearings as the "true" cause of injuries. Since most older women are menopausal and very many younger women have had recent pregnancies, it seems to be rather easy to attribute women's MSDs to personal rather than occupational factors. (Messing 1998, 97–98)

Actual empirical studies of the relationships between women, work, and health are few and far between, Messing argues. As with pharmaceuticals, women are routinely excluded from occupational health studies and are then assumed to be the same as men.

> Women have been eliminated from studies on the ground that their problems have not been demonstrated. One million-dollar government-supported study related cancers to a huge number of occupational exposures. When we asked the researcher why his study excluded women, he replied, "It's a cost-benefit analysis; women don't get many occupational cancers." He did not react when we suggested that his argument was circular. (1998, 62)

In his book, *Occupation and Disease,* Allard E. Dembe described how women's strain injuries have been treated quite differently from men's strain injuries, in law and by medical practitioners. "In the early twentieth century, telegraphists' cramp became perhaps the first chronic disorder caused by physical hazards to be specifically deemed compensable under workers' compensation laws," he wrote. But a few years later, the very Industrial Disease Committee that recognized telegraphers' cramp, rejected "twisters' cramp," which affected mostly women lacemakers (Dembe 1996, 41).

Terminology inadvertently reinforced the divisions caused by stereotyping. Pre-Freud, conditions affecting the nerves were commonly called "neuroses." If caused by work, they were called "occupational neuroses." When the budding field of psychology adopted the word "neurosis" to label mental disorders, "occupational neuroses" slid into the category of anxiety disorders. "A temperamental predisposition to nervous excitability came to be recognized by physicians as a major factor in the development of occupational neuroses," Dembe noted. "In the medical and popular literature of this period, two types of people were singled out as possessing these traits in a disproportionately high degree: females and Jews" (Dembe 1996, 49).

These stereotypes have persisted and been revived in following epidemics. Although some medical practitioners and researchers recognized that women's paid and unpaid work was as demanding and intense as men's paid work, the most influential did not. Take Dr. George Phalen, who in the 1950s pioneered and perfected the surgical carpal tunnel release still widely used today. Dembe writes: "Phalen observed that most of the patients who came to his clinic were women, many of whom were in their middle years (aged thirty to sixty), and most of whom did not work in traditional industrial settings. He concluded that manual work could not be a cause of the disorder because, to him, these women did no 'manual' work. But the cases tell another story" (Dembe 1996, 71).

Similar blinders seemed to influence doctors who dealt with the Australian RSI epidemic in the mid-1980s. Dr. Damian Ireland expressed a common point of view in an article published in a 1986 issue of *Australian Family Physician*. "RSI is best defined as an occupational neurosis affecting young to middle aged (predominantly female) employees engaged in the low paying, monotonous 'low glamour' occupations," he wrote. His recommended course of treatment was psychotherapy, but first "the patient must be prepared to accept that some component of the condition is psychological. An aggressive denial at this stage indicates poor prognosis and renders subsequent treatment ineffective" (Ireland 1986).

By contrast, Statistics Canada stated unequivocally that the differences in women's and men's rates of strain injuries, "are likely attributable to the types of activities each sex undertakes. . . . Just over half of RSIs sustained by men and women happened while working. For men, sports or physical exercise was the next most frequently cited activity, whereas for women, activities relating to chores, unpaid work or school ranked second" (Statistics Canada 2003).

The AFL-CIO (n.d.) also makes the link between activities and injury:

> "MSDs [musculoskeletal disorders] result from working conditions such as a fast pace, heavy lifting, repetition or working in an awkward and uncomfortable position. These injuries are common in a wide variety of occupations including nursing aides, cashiers, poultry workers and sewing machine operators. Those occupations most affected by MSDs also tend to be lower wage jobs that employ high numbers of minority workers."

Karen Messing makes the point that few researchers even know exactly what work women's jobs involve:

> Women's segregated work has its own physical demands, not usually recognized in the workplace. . . . We recently interviewed a group of male and female hospital cleaners about the physical difficulties encountered in their work. The women complained to us that women who wish to be assigned to "heavy" cleaning (now 100 percent male) must take a strength test, while

> men who wish to be assigned to "light" cleaning (now 80 percent female) are not tested in any way. We observed the physical requirements of "light" cleaning, which involved constant bending and stooping, picking up and dusting small objects, very quick scrubbing movements with the arms and hands, and nearly constant motion over a long work day.
>
> We also observed the physical requirements of the men's jobs: slower and somewhat less frequent movements, often against greater resistance, usually carried out while walking, with more rest breaks. The physical requirements of both jobs were considerable, but quite different. Only one of the two types of jobs was considered to require strength that could be tested by a pre-employment test. (1998, 37)

Seventy percent of women in the Canadian workforce are still congregated in so-called traditional occupations for women: teaching, nursing and health care, clerical work, and service industries. Two-thirds of all minimum-wage workers are women, as are nearly 70 percent of part-time workers, many of whom work two or more part-time jobs in order to make ends meet. A scant 25 percent of senior management people in Canada were women in 2002 (Statistics Canada 2003).

"Work stress deriving from a fast work pace, role ambiguity, worry and monotonous tasks has been associated with RSIs in the past," according to the StatsCan report, and that association was reinforced by this survey. However, stress is not the only factor. The worker's level in the workplace plays a bigger role, with low-paid, low-prestige workers having the highest rate, as a rule. "Least likely to be injured," StatsCan concluded, "were people in management" (Statistics Canada 2000).

Among the advantages accruing to people in management is that they are more likely than nonmanagement to be able to adjust their workstations to fit them. One-size-fits-all workstations are particularly hard on women. "A traditional work surface in the United States is 29 to 31 inches [74 to 79 cm] high," according to Dr. Linda H. Morse and nurse practitioner Lynn J. Hinds, in an article published in *Occupational Medicine: State of the Art Reviews.* This height is "ideal for a man who is 5 feet 10 inches [178 cm] tall but totally inappropriate for a woman of 5 feet 1 inch [155 cm], who requires a work surface approximately 23 to 25 inches [58 to 64 cm] in height" (Morse and Hinds 1993).

Since World War II, anthropometric measurements literally have been "measured for men." Tables, chairs, desks, conveyor belts, hand tools, and workstations are all designed to accommodate the so-called average man, who is assumed to be the same size as the average U.S. army recruit during the war.

Morse and Hinds argue that ergonomists need to pay more attention to the changing workforce, which includes more and more women as well as members of ethnic minorities (such as Hispanic and Asian people) who tend to be smaller than 5 feet 10 inches [178 cm]. "Failure to do so," they warn, "will perpetuate the rising rate of injuries due to ergonomic hazards" (Morse and Hinds 1993).

Of course, governments can encourage employers to protect their workers, by incorporating ergonomics requirements into labor standards. Three provinces in Canada have passed ergonomics regulations: British Columbia, Alberta, and Saskatchewan. Alberta seems to have rescinded its ergonomics requirement in the 2003 revision of the Occupational Health and Safety Act, however. The British Columbia Occupational Health and Safety Regulation "General Conditions" section says, "The employer must eliminate or, if that is not practicable, minimize the risk of MSI to workers." The Saskatchewan Ministry of Labor's Web site says that "although the term ergonomics is not specifically used in The Occupational Health and Safety Regulations, 1996, the topic is dealt with in sections 78, 79, 80, and 81 of the regulations." These sections are concerned with:

1. lifting and handling loads;
2. work that involves standing for long periods of time, anti-fatigue mats and footrests;
3. situations where it is appropriate for workers to be permitted to sit while working (even if historically the job has been done while standing), seating requirements and footrests;
4. musculoskeletal injuries which include muscle injuries or disorders of tendons, ligaments, nerves, joints, bones, etc. (Ministry of Labour 2005)

In July 2003, Manitoba Labour and Immigration introduced draft legislation that would require employers to control MSD hazards in the workplace (www.gov.mb.ca/labour/safety/regreview/pdf/ergonomics.pdf).

In addition, the federal government amended the Canada Labour Code in 2000 to include mention of an ergonomics standard. Under Part II, Duties of Employers, section (u) requires that employers "ensure that the work place, work spaces and procedures meet prescribed ergonomic standards" (Canada Labour Code). Unfortunately, despite its name, the Canada Labour Code applies to only about 10 percent of the workforce—those industries that come under federal jurisdiction, such as banking, maritime shipping, interprovincial transportation, and telecommunications.

On the voluntary side, in 2000 the CSA (formerly the Canadian Standards

Association) released CSA-Z412–00, a comprehensive office ergonomics standard that addresses office issues including job duties, building ventilation, hours of work, lighting, and computer workstations (Canadian Standards Association 2001). In July 2002, the American National Standards Institute (ANSI) accredited CSA to develop standards for built environments such as office buildings, including ergonomic standards.

Accrediting subcontractors seems to be one way that ANSI can approve ergonomics guidelines. A proposed update to the 1988 ANSI computer workstations guideline was stuck in committee for more than a decade. Finally, in September 2002, ANSI announced that the Human Factors and Ergonomics Society was releasing a draft computer workstation guideline, in its role as an ANSI-accredited standards developer (American National Standards Institute 2002). ANSI stated that the HFES document (BSR/HFES100, *Human Factors Engineering of Computer Workstations* ("guides designers on how to accommodate variation in both the size of individual users and the manner of usage" (Human Factors and Ergonomics Society 2002).

While there has been some resistance to ergonomics regulations, particularly in British Columbia, for the most part Canadian legislative and voluntary ergonomics standards have been adopted without the sort of major political battles that have raged in the United States. And while occupational health and safety issues suffer some of the same stigma in Canada as in the United States, unions are stronger in Canada (representing about 30 percent of employed men and the same percentage of employed women) and provide a larger population base for political action.

Canada also has a social-democrat political party called the New Democratic Party (NDP), which elects representatives to the federal Parliament and most provincial governments. The NDP often brings working peoples' concerns to the House of Commons. In 1999, for example, an NDP member of parliament brought a private bill to the House to amend the criminal code so that management of a corporation could have been found guilty of what is known in many other countries as "corporate manslaughter." The governing Liberals persuaded her to withdraw the bill so that they could propose their own act, which came into effect on March 3, 2004 (United Steel Workers of America 2003).

One political activity that originated in Canada and spread around the world is the annual April 28 International Day of Mourning for Persons Killed or Injured in the Workplace, called Workers Memorial Day in the United States. That use of the word "Persons" is peculiarly Canadian, evoking a 1929 decision by the British Privy Council (then the highest court in the Commonwealth) that women are indeed "Persons" for the purposes of Canadian legislation.

In a similar vein, Torontonian Catherine Fenech founded International RSI Awareness Day in 1999, which has since been picked up by some major unions in twelve countries, including Australia, the United States, the United Kingdom, as well as Canada (International RSI Awareness Day n.d.).

There is one huge advantage to being injured in Canada rather than, say, the United States. Canada (like most of the industrial world) has universal health care insurance. Injured Americans sometimes say that they are forced to remain at the jobs that hurt them because that is the only way they can afford the health care they need. Their insurance is tied to their employment. In contrast, said Catherine Fenech:

> Universal health care, as we have in Canada, provides the greatest choice. If I don't like the opinion of one doctor, I can get a second, third, fourth opinion and so on. I can see as many doctors as I want. There is no limit to who I can see. I also don't have to worry about going bankrupt because I have the misfortune of getting sick or being hospitalized. If I lived in the US, I would have been in dire straights with my injury. I got hurt working part time so I would have had no health insurance (I have no extended health care coverage here). My WC [workers' compensation] claim was initially denied so I wouldn't have had health coverage there. So where would I be? Well, fortunately, I live in Canada.
>
> I was able to see a doctor immediately, be referred to specialists, receive the necessary tests, go to physiotherapy, relaxation therapy, and even Feldenkrais lessons without paying a cent out of my pocket. This was all covered by my provincial health insurance, which every resident has a right to. Without universal health care, I would have had to wait for 9 months for my WC claim to be allowed before receiving health care. Although I guess my claim would not have been allowed because I wouldn't have had the medical documentation to support the claim. (1999)

Although Catherine Fenech had access to medical care, her WC claim dragged on for five years, from 1995 to 2000. Interestingly, she said that the experience politicized her. Before her injury, she described herself as "definitely conservative." As her union supported her WC claim, she moved to "left of centre (not too far though)" (Kome n.d.). If the old saw is correct that a conservative is a liberal who has been mugged, then perhaps a liberal is a conservative who has gotten injured at work. Canada is a slightly more liberal and socially aware society than the United States in many areas besides health care: strict gun laws; lower tuition fees for post-secondary education; more acceptance that immigrants can keep their own diverse cultures; a "child tax benefit" that delivers monthly payments to most mothers while their children live at home; and of course, the Canadian equivalent of the Equal Rights Amendment entrenched in the 1982 Charter of Rights and Freedoms.

In addition, Canada has been the setting for a concerted union-led drive to improve occupational health and safety regulations. In his book *Consulted to Death,* Doug Smith recounts how Bob Sass led a national campaign for the "Three R's": the right to know about workplace hazards, the right to be participate in workplace safety decisions, and the right to refuse dangerous work. "While the central goal of these reforms was to reduce the severity and frequency of accidents and reduce occupational disease," Smith wrote, "Sass and others hoped that they would serve to democratize the workplace and to remove health and safety from marketplace regulation. . . . Just as it is illegal to sell oneself into slavery in Canada, it should not be possible to sell one's health." (Smith 2000, 43).

Smith covered Bob Sass's Occupational Health and Safety campaign in the 1970s, and returned in the late 1990s to evaluate whether the campaign was successful. His rueful conclusion: "One hundred years ago workers, not work organization, were seen as the main cause of all on-the-job injuries. And just as was the case one hundred years ago, there is now little interest in addressing health as opposed to safety issues." (Smith 2000, 140).

Conclusion

To sum up: workers seeking help with strain injuries in Canada may find their situation a bit less desperate than similar workers in the United States. All the same, the Statistics Canada Community Health Report presents evidence that the number of workers with strain injuries continues to increase, despite current BLS and AWCBC figures. In Canada, unlike the United States, injured workers (and especially injured women) have the consolation of being counted and having access to affordable, quality health care.

Canada's population is roughly 30 million, or approximately ten percent of the U.S. population. Americans might be interested to know that, before the federal government founded Statistics Canada in 1985, most Canadians created their own statistics by finding the relevant U.S. data and moving the decimal point one place to the left—making 10 percent. Conversely, if the StatsCan data are correct, that means that 30 million Americans are living with serious strain injuries—as many as the total population of Canada.

References

AFL-CIO. n.d. "Women Workers Need an Ergo Standard" factsheet. Available at: www.aflcio.org/yourjobeconomy/safety/ergo/ergowomen.cfm.

American National Standards Institute (ANSI). 2002. "Voluntary Standards Cover the Spectrum: From Computer Workstations to Lasers." ANSI news release September 20. Available at: www.ansi.org/news_publications/news_story.aspx?menuid=7&articleid=272.
Association of Workers' Compensation Boards of Canada (AWCBC). 2004. Available at: www.awcbc.org.
Canada Labour Code. 2003. Part II: Occupational Health and Safety. Available at: http://laws.justice.gc.ca/en/1–2/17243.html.
Canadian Standards Association. 2001. "CSA Z412-Guideline On Office Ergonomics." May 17. Available at: www.csa.ca/products/occupational/default.asp?.
Dembe, Allard E. 1996. *Occupation and Disease: How Social Factors Affect the Conception of Work-Related Disorders.* New Haven, Yale University Press.
Fenech, Catherine. 1999. Sorehand post September 14. Available at: http://sorehand.org/archives/1999/msg03574.html.
Human Factors and Ergonomics Society. 2002. BSR/HFES100, *Human Factors Engineering of Computer Workstations*, March 31. Available at: www.hfes.org/publications/hfes100.html.
International RSI Awareness Day. n.d. Available at: www.ctdrn.org/rsiday/.
Ireland, Damian C. 1986. "Repetitive Strain Injury." *Australian Family Physician* 15, no 4 (April)
Juster, Thomas F., Hiromi Ono, and Frank Stafford. 2002. "Time-use Studies. Available at: www.umich.edu/~newsinfo/Releases/2002/Mar02/r031202a.html.
Kome, Penney. n.d. Private correspondence with Catherine Fenech.
Messing, Karen. 1998. *One-Eyed Science: Occupational Health and Women Workers.* Philadelphia: Temple University Press.
———, ed. 1999. *Integrating Gender in Ergonomic Analysis: Strategies for Transforming Women's Work.* Brussels, Belgium: European Trade Union Technical Bureau for Health and Safety.
Ministry of Labour. 2005. "Occupational Health and Safety. Ergonomics." Saskatchewan. Available at: www.labour.gov.sk.ca/safety/fast/ergonomics.htm.
Morse, Linda H., and Lynn J. Hinds. 1993. "Women and Ergonomics." *Occupational Medicine: State of the Art Reviews* 8, no. 4 (October–December).
Occupational Safety and Health Administration. 2003. "OSHA Issues Final Rule on Recordkeeping Form," June 30. Available at: www.osha.gov/pls/oshaweb/owadisp.show_document?p_table=NEWS_RELEASES&p_id=10281.
O'Neill, Rory. 1999. *Europe Under Strain.* Brussels, Belgium: European Trade Union Technical Bureau for Health and Safety.
Ramos, Xavier. 2003. "Domestic Work Time and Gender Differentials in Great Britain 1992–1998: Facts, Value Judgements and Subjective Fairness Perceptions." Paper presented at the British Household Panel Survey 2003 conference at the Institute for Social and Economic Research (ISER), University of Essex.
Smith, Doug. 2000. *Consulted to Death: How Canada's Workplace Health and Safety System Fails Workers.* Winnipeg, Canada: Arbeiter Ring Publishing.
Statistics Canada. 2000. *The Daily,* September 14. Available at: www.statcan.ca/Daily/English/000914/d000914c.htm.
———. 2003. *The Daily,* August 12. Available at: www.statcan.ca/Daily/English/030812/d030812b.htm.
U.S. Fedstats. 2004. Available at: www.fedstats.gov.

United Steel Workers of America. 2003. "Steelworkers Applaud Passage of Corporate Responsibility Act," October 30. Available at: www.uswa.ca/eng/news_releases/westray_leg.htm. (n.d.). Available at: www.hazards.org/women/understrain.htm.

"Women's Health and Safety: Women's Work, Women's Burden." n.d. *Hazards*. Available at: www.hazards.org/women/understrain.htm.

Workplace Safety Insurance Board (WSIB). 2004. Toronto, Ontario. Available at: www.wsib.on.ca/wsib/wsibsite.nsf/public/home_3.

9

All That Is Solid Melts Into Air

Worker Participation and Occupational Health and Safety Regulation in Ontario, 1970–2000

Robert Storey and *Eric Tucker*

Over the course of the past twenty years governments in most advanced industrial capitalist societies have altered the dynamics, if not the actual statutes and regulations, of their occupational health and safety (OHS) regulatory regimes. In one form or another, and to differing degrees, these changes represent a shift away from systems of mandated partial self-regulation that consisted of an internal responsibility system in which workers enjoyed participatory rights, supported by external enforcement, toward more ambiguous paradigms characterized by a conscious downsizing of government intervention and a heightened emphasis on employer self-regulation without a concomitant commitment to worker participation. Government activities in the realm of OHS are thus being brought in line with state initiatives in other social, economic, and political arenas where programs and policies associated with greater worker and citizen rights have been systematically undermined and/or eliminated altogether. In other words, the shift in the cosmology of OHS regulation from a philosophy of ethical self-compliance to a decided emphasis on self-regulation is more than an issue of semantics. It is, rather, a signal that the politics of neoliberalism—most particularly policies and programs associated with deregulation—are now an integral part of the political economy of occupational health and safety.

The reasons for this move toward greater self-regulation and, in effect, the further development and consolidation of state-sanctioned management health and safety systems are well known—at least from the perspectives of business and government. According to officials from both groups, the opening of national economies to global competition has necessitated the reshaping of those economic and policy frameworks that facilitate the competitive production and distribution of goods and services. With respect to OHS, the response to these "global" pressures has been a move toward regulatory processes that allow companies to produce without the "burdens" of having to comply with "excessive" OHS standards while actively promoting a primary role for management in company-based health and safety systems.

In Ontario, the longstanding industrial and financial heartland of Canada, the smoldering forces of neoliberalism burst into flames in the mid-1990s when a newly elected Progressive Conservative Party (PC) government embarked on a program of dismantling and/or refashioning key components of the province's economic, social, and political infrastructure. With respect to the concerns of this chapter, one of its first initiatives was reversing key facets of collective bargaining legislation that had been the (controversial) landmark of the outgoing social democratic New Democratic Party (NDP) government. This action outraged Ontario unions who promptly organized "Days of Protest" in cities across the province. While two of these protests, one in the industrial steel center of Hamilton and the other in the provincial capital of Toronto, each brought close to 100,000 people into the streets, they had little immediate impact on the PC government and its neoliberal agenda. Indeed, it was in the midst of these protests that the government turned its attention to revamping an OHS regulatory regime that had been twenty years in the making. In specific terms, it took aim at dismembering a number of OHS educational and regulatory bodies that the business community in the late 1980s had targeted as both too costly and as encroaching too far into the area of management rights. As it happened, however, they were not completely successful in these endeavors. After cutting funding to the Ministry of Labour, shutting down the Workplace Health and Safety Agency (WHSA), an embattled bipartite body responsible for OHS educational and training, and refusing to act on recommendations to lower exposure limits to hundreds of real and potentially dangerous chemicals and substances, further plans to reduce the number of OHS inspectors and to mothball a system of occupational health clinics run and managed by the labor movement were thwarted by sustained protests from OHS union activists (Walker 1997).

Notwithstanding this—and other—important moments of successful opposition to the neoliberal agenda of Ontario business and the PC government, it is clear that over the course of the past decade the central position that OHS once held within the political economy of provincial politics and the labor movement has greatly diminished. The advances made by workers and unions in the course of struggles throughout the 1970s and 1980s—struggles that can be captured in the phrase "writing the workers in"—have either been turned back via government legislation or made relatively benign by ascendant corporate OHS agendas that leave little or no room for worker/union participation. Workers are, in short, being "written out" of OHS regulatory policies and programs. Further, and in direct relation to the above, the OHS movement is in a state of disarray. In place of a mobilized and politically vibrant OHS movement, OHS activists were isolated from one another and on difficult and uncertain organizing terrain. On one hand, like union activists more generally, they operate in a climate where the interests of capital are dominant and job insecurity is high—fertile grounds for the seeding and nurturing of job fear and paralysis. On the other hand, they also confront a unique set of internal barriers in the form of publicly funded, labor-managed, educational/training institutions that are seen by large numbers of these activists as serving the dysfunction of underwriting the related processes of depoliticizing and demobilizing the OHS movement. All that was solid is melting into air.

This chapter offers an analysis of the evolution of OHS regulation in Ontario from the major reforms of the 1970s to the present. The argument that will be presented is that although there has been little formal change in the major parameters of OHS legislation in Ontario from the early 1980s, the rise of neoliberalism in Ontario has nevertheless transformed the regime's operation. In presenting this argument, we will first outline and analyze the developments and factors that led to the formulation and passage of an OHS act that statutorily enshrined workers' rights to know, participate in, and refuse unsafe work and provided for more effective and comprehensive direct state regulation. The second section will examine how the changing political economy has affected the labor movement generally and the health and safety movement in particular to explain why a better institutionalized role for firm-level worker participation and provincial-level union participation has been accompanied by a reduced level of worker health and safety activism. Third, we will explore how neoliberalism has reshaped governmental approaches to OHS regulation. In particular, we will highlight how the emphasis on self-reliance and OHS management systems in the context of a weakened labor movement threatens worker health and safety. We will also take account of shifting patterns in the production and enforcement of public health and safety standards. As stated above, it will be argued that these developments point to an important shift in the regulatory framework in terms of worker participation: from being written in to being written out. Finally, we will discuss how these alterations have precipitated a critical rethinking of strategy among OHS activists in the labor movement.

Writing in the Workers

The 1970s and 1980s were heady days for OHS activism in most advanced capitalist societies.[1] During these two decades workers and unions in Canada, the United Kingdom, Sweden, Australia, and the United States, to name only a few, were successful in their efforts to have their respective governments pass OHS laws that established, albeit in varying degrees, a legislative framework that statutorily enshrined workers' rights to participate in maintaining and improving the health and safety of their workplaces and strengthened direct government regulation.[2] In the views of OHS activists these new laws were important for two reasons. First, they constituted a set of rules and regulations that regardless of their limitations constituted a definitive advance over what had been in place. Second, the participatory rights gained by workers seemed unparalleled in their encroachment on management prerogatives and, as such, were perceived by many as a central pathway to advancing and deepening the structures and processes associated with industrial democracy and citizenship at work.

In Ontario, the comparable changes and advances came in the late 1970s and stemmed primarily from strong pressures exerted on the government by a new and vibrant OHS movement (Storey 2004a). No doubt inspired by the successful struggles of American hard rock miners for compensation for black lung disease and by advances in OHS legislation in the province of Saskatchewan, the fledgling OHS movement stirred first among nickel miners in northern Ontario in the late 1960s. However, the movement took shape and direction in the face of revelations in the mid-1970s that uranium miners in the small northern town of Elliot Lake and employees at a Johns-Manville pipe insulation plant in a suburb of Toronto were experiencing elevated and deadly rates of disease and cancer. These events spurred both the labor movement and its political ally, the NDP, to demand that the government enact legislation that would both protect workers and allow them an active role in ensuring their health and safety (Sass 1989).

Passed by the Ontario legislature in the fall of 1978, Bill 70, "An Act Respecting the Occupational Health and Safety of Ontario Workers," gave Ontario workers the right to know about the substances they worked with, to form joint health and safety committees (JHSC) and inspect their workplaces, and to refuse work they believed was unsafe. It also provided for more effective regulation, particularly in respect to health hazards. While a definite step beyond existing legislation, many OHS activists nevertheless saw Bill 70 as a compromise measure. According to D'Arcy Martin, then an educator with United Steelworkers of America (USWA), the OHS Act was "not what the militants wanted; it is what the militants settled for." In place of a system that provided workers and unions with central roles in the regulatory process, the new act gave them "a vague right to know, a right to participate via JHSCs that were only advisory, and rights to refuse unsafe work that defined it as an individual right. There was no right to a collective refusal as had been demanded by OHS militants" (Martin interview).[3]

From the outset, the regulatory scheme contained two tensions. The first was over the strength of worker rights in what became known as the internal responsibility system (IRS). The second was over the roll of direct state regulation and enforcement. As it evolved, workers/unions, employers, and the state had starkly different visions of how the system should operate. Employers, except for a small group who had previously been involved in formal health and safety arrangements with (mainly) unionized workforces, proved to be extremely reluctant to institutionalize internal responsibility (O'Grady 2000; Lewchuk et al. 1996). They were upset with the provisions in the act that obliged them to provide a safe and healthy working environment, fearing they would be "burdened" by a mounting array of obstructive and unattainable standards and regulations. They saved their strongest critique, however, for the right of workers to refuse unsafe work. From the moment this provision was included in discussions to its appearance in the proposed statute, employers steadfastly opposed it, arguing that workers would abuse the right, and, perhaps more significantly, that such actions posed a direct challenge to their right to manage (Walters 1991).

From the outset, government adopted the view that its primary role was to support the IRS. That is, while government inspectors were legislatively in place to monitor and police the act, enforcement actions were only to be taken when all else had failed.[4] Indeed, it was the perception of many activists and trade union officials at the time that the OSH Act became a means for the government—under siege from the NDP and OHS activists for most of the decade —to relieve itself of this pressure and, ultimately, "abdicate its responsibility to workers" (Martel 1995; Edwards interview).

The sole enthusiasts in the new configuration were the OHS activists and the labor movement. Indeed, while many of the "militants" in the health and safety community were disappointed and angry about the "compromises" contained in the act, they came to realize that Bill 70 could serve as a "critical factor in giving [workers] space" (Martin interview). Dan Ublansky, now a lawyer with the Oil and Chemical Workers Union, remembers the excitement of these times and the potential for real change many believed was imminent.

> In '79 there was euphoria around having achieved this tremendous legislative change, victory. . . . It seemed like such a departure from what had

> been the norm before that. It opened up a whole new area for activism. . . . Some of it was already here through collective bargaining, but the idea of having workers involved in health and safety and having some say in how things are done and the right to refuse had this sort of mystical quality about it. That, you know, we did all the talk about power, and empowerment. . . . That knowledge is power. That was the message that was being delivered. So people believed it and that created a lot of energy and enthusiasm. . . . We were hungry for knowledge. Let's get all the knowledge that we can and then we're going to use this knowledge and convert it into power by making big changes in the way things are done. (Ublansky interview)[5]

The 1980s were, indeed, a period of intense activism. According to Ublansky, by the "mid-1980s occupational health and safety was one big battleground" with the major conflicts centering on the real and perceived limitations of the JHSC, the cautious role of inspectors, and the failure of the governments of the day to actively enforce the act (Ontario NDP 1983; 1986).[6] For Clarence Macpherson, former worker at a Chrysler assembly plant, but in this period an employee of the Ontario Federation of Labour's (OFL) OHS Department, these contests turned on the realization by workers and OHS activists that the "Three Rs" were little more than "paper rights."

> As workers [took] on these additional rights and responsibilities under the Health and Safety Act [it became] clear that some of them [were] purely paper rights. Like the right to know! How can you really have a right to know if you don't have a regulation that stipulates what chemicals you're handling, that identify them and give you access to data sheets and what have you. . . . And then you have deficiencies . . . around the right to refuse. Interpretive bulletins, etc., that the ministry put out that would limit the effectiveness of the worker's right to refuse dangerous work. . . . So, those became contentious issues and required clarity either in policy within the Ministry [of Labour] or in the form of regulations or changes to the act that would better define the right to refuse. And then the activists [who] were being trained through the '80s . . . [who] through their lobbying efforts with the government were pressing for expanded rights and responsibilities. (Macpherson interview)

Throughout the 1980s workers and unions put these key provisions of the act to an active test. The most significant of these actions occurred in the latter part of the decade when thousands of workers at de Havilland and McDonnell Douglas, aerospace plants on the edges of Toronto, went out on strike over health and safety conditions at their plants (Livesey 1988). Of the two events, the strike at McDonnell Douglas was the more dramatic. Here, workers became increasingly upset over the company's unwillingness to address a report of OHS conditions with McDonnell Douglas that had itemized hundreds of violations of the OHS Act. What finally pushed the workers into action, however, was a lack of action on the part of the government—even after Ministry of Labour inspectors validated the findings of the union-sponsored report. As Nick DeCarlo, then local Canadian Auto Workers (CAW) union president at McDonnell Douglas, recalls:

> So we had a meeting with the Ministry of Labour and . . . the comment they made was, well, if its so bad how come your workers aren't refusing? OK. . . . The Ministry of Labour says if things are so bad why aren't your workers refusing, right. So, in a short time we had 2,500 work refusals. . . . It lasted from mid-November to about Christmas. . . . The position it put the company in was every work refusal had to be investigated by the health and safety rep. We had 2,500 work refusals and each one has to be investigated and documented and you know it can take several hours to work on it. It would take forever to do 2,500 work refusals, never mind resolve them. Never mind bringing an inspector in for each one. So they were in a position where they were forced to negotiate. . . . We finally resolved a number of issues. . . . [W]e got ventilation issues resolved and we got a whole bunch of health and safety changes, etc. So, by and large it was a major victory. (DeCarlo interview)

While a "counterattack" by the company in the form of layoffs and an announcement that "some of the work was being moved out of the plant" took the steam out of the workers, two points about these strikes need to be underscored. First, the specter of thousands of workers at these plants challenging the law by engaging in a mass work refusal was illustrative of a newfound consciousness among these workers, and workers in Ontario more generally, that their "health was not for sale."[7] As such, it highlighted both the willingness of workers to back their beliefs with action and their sense that collective action on OHS issues could, and did, lead to greater worker influence and power in the workplace. According to DeCarlo, after this historic and arguably "illegal" strike over health and safety conditions at McDonnell Douglas, "the whole attitude of the plant was quite dramatically different than it had been before that. . . . You had a period of about five or six weeks where people kind of felt they ran the plant and effectively some of the workers did." Second, these events had wider political applications. In short order, the Liberal Party government (the PC government had been defeated in 1985) implemented a strong version of the national Workplace Hazardous Materials Information System (WHMIS) relating to the labeling of and provision of data to workers on hazardous products found in workplaces and the mandatory OHS education and training for all workers, agreed to the establishment of two government-funded, labor-movement-managed, occupational health clinics, and fueled deliberations regarding major changes to the OHS Act.

In the midst of this whirlwind of political and legislative activity, OHS activists began to feel a rising tide of resistance from Ontario employers. After having been defeated in their attempts to revamp the workers' compensation system, cajoled into participating on a joint management-labor committee whose mandate was to develop regulations lowering exposure levels to hazardous substances, and watching the creation of the two occupational health clinics, they had reached their limits and were poised to strike back.[8] Hence, along with maintaining ongoing subtle forms of resistance, such as neglecting to establish joint health and safety committees and disciplining workers whom they deemed to be exercising the right to refuse in abusive ways (Harcourt and Harcourt 2000; Walters 1991), Ontario employers from all sectors embarked on a more overt defense/assertion of their perceived rights and interests. This increasingly vehement stance became starkly evident in government hearings to discuss proposed Liberal government amendments to the OHS Act that extended coverage to tens of thousands of office workers, gave workers the unilateral right to stop work, and laid the legislative groundwork for the creation of the WHSA to be administered jointly by labor and business. In these hearings, employers largely ignored the proposed inclusion of white-collar workers, choosing instead to focus their attacks on the unilateral right to stop work and the WHSA. With regard to the former, employers relayed their strong belief that the current provision giving individual workers the right to refuse unsafe work had worked well and was thus all that was required in this area. In this vein, Linda Ganong, the director of provincial affairs for the Canadian Federation of Independent Business (CFIB), told the Standing Committee on Resources Development that her organization was "relentlessly opposed to the unilateral power to stop work. Unilateralism is completely contrary to the first principles of health and safety management. It will corrode the trust between employers and employees, which is essential to the achievement of health and safety goals. It will seriously impair the functioning of the internal responsibility system" (CFIB 1990, R-5). Barbara Caldwell, chair of the Ontario division of the Canadian Manufacturer's Association (1990, R-117), was equally adamant in voicing her association's opposition to this provision.

> Speaking first of all to the stop work provisions, in an emergency in our operations, any employer can shut down our equipment and is expected to do so. In fact, it is part of his or her job description. But shutting down a

> plant is not a trivial matter, particularly in complex continuous operations such as, for example, a nylon polymerization unit such as we have in Kingston. It could leave the unit down for days, cost millions of dollars and, in itself, cause major safety problems. Such a shutdown would not occur trivially. It would be discussed and assessed by management, by engineers, by technical people and by operators before such a decision was made, except in an emergency. It would be impossible for us or anyone else to train a worker representative to weigh and assess all the aspects of such a shutdown.

Turning to the WHSA, employer groups had three principle concerns: that the labor representatives on its board of directors were to be drawn solely from the labor movement; that the mandate of the agency threatened to usurp the work and role of independent (i.e., employer) safety associations; and, that in an era of heightened competition, fulfilling the mandate of the WHSA (i.e., training certified worker and management OHS representatives for all workplaces covered by the act) would prove to be too costly.[9] Yet, ultimately, their concerns were larger than specific fears that a unilateral right to refuse would be abused or that the WHSA would be labor dominated and expensive. Rather, Ontario employers saw Bill 208 as expanding the rights and powers of workers and unions in the regulation of OHS. It was writing workers further into the regulatory framework that they both feared and opposed.

In these developments, then, we see the first stirrings of *organized* employer opposition to the agenda and vision of OHS activists within the Ontario labor movement.[10] As we will see below, these nascent organizational formations solidified in the early 1990s as part of a generalized opposition by employers to what they understood to be the excesses of the Canadian welfare state in general, and to the specific policies and programs of the Ontario NDP, which came to power in the provincial elections of 1990. At the same time, however, it is also critical to point out that employer opposition was not the only obstacle in the path of OHS activists and further OHS reform. During this period another roadblock was being laid by influential union leaders who were growing progressively uneasy about the impact that years of activism and radical critique were having on their memberships. To return again to the example of de Havilland and McDonnell Douglas, Cathy Walker, an OHS activist from British Columbia who became director of health and safety for the CAW in the early 1990s, recalls how these episodes served to energize the rank and file to examine all aspects of their workplace lives. According to Walker, one facet of this wholesale review concerned the processes through which CAW members became part of the union's staff. Seeing that many staff positions in the union were filled by appointment rather than elections, the demand was raised that this be changed. Shop floor activism was spilling into the union hall and this shook up the leadership.

> I think people [were] a bit terrified by the McDonnell Douglas, de Havilland thing. I think that people had traditionally seen health and safety as much more a collaborative, co-operative kind of thing. You know, giving old Joe the job in the plant and he can run around and effectively help out the boss improve the place and tell people to wear their safety boots and see the role as a health and safety cop. (Walker interview)

The CAW was not the only union to feel the pressure from OHS activists within its ranks. National and international leaders within the USWA also became concerned with the direction of OHS politics in their union. After leading the struggles over occupational disease at Elliot Lake and the Steel Company of Canada in Hamilton, Ontario, in the 1970s, the USWA was the recognized leader in OHS among unions in the province. This leadership was fueled by a militancy at the rank-and-file level that overfilled its container when Dave Patterson, a combative local union president representing thousands of Inco miners, won a 1981 election for USWA district director over a candidate favored by the Canadian and international leadership of the union. In turn, Patterson staffed his office with individuals such as Elliot Lake–veteran Ed Vance, whom D'Arcy Martin recalls was "a militant rank-and-file leader who was passionate about this stuff and with a sense of obligation to his dead comrades. He would not back down on anything." However, as Martin also recounts, the promotion and/or willing tolerance of OHS militancy within the upper echelons of the USWA changed dramatically "when Patterson was defeated [in 1985]. The lid was put back on. . . . I am not making a moral judgment on what the people did. The culture and the possibilities of health and safety mobilization did narrow after the Patterson defeat. . . . It is not like it went away. But some of the bite went out of it."

A desire to narrow or channel the activities of the OHS activists was only part of the dilemma confronting leaders within the Ontario labor movement. They also had to decide where to direct that energy and commitment.[11] Their answer was bipartism. As understood by these union leaders, bipartism involved trade unions and trade union officials taking their rightful place alongside employers in decision-making processes that affected the economic and physical well-being of Ontario workers. Bipartism was not, however, a new issue within the ranks of Ontario unions; indeed, its adoption as an official labor movement position had been discussed and defeated at previous OFL conventions. In those bitter and heated debates, opponents of bipartism warned that despite formal parity on committees, and the like, workers and unions would always come out second best. Worse still, the critics argued, trade union participation in bipartist committees and negotiations would ultimately co-opt the labor movement as it would become complicit in legitimating and enforcing any decisions unfavorable to workers.

As it happened, Ontario's union leaders found unlikely allies in their push toward bipartism within the ranks of their former opponents. That is, many OHS activists, after being involved in health and safety struggles for a decade or more, were beginning to believe that continued activism and militancy were unlikely to prompt further change—either from employers or the government. Dan Ublansky, for example, reluctantly embraced the shift toward bipartism as "nothing [else] was working so it seemed like a good idea." More importantly, Ublansky and other former critics supported a bipartist position in OHS because "the labour movement thought it was smarter than anyone else."

> It seemed like a good idea at the time because after ten years or eight years of conflict and not much in the way of concrete results, looking forward to just carrying on the fight didn't seem like a particularly smart strategy. It hadn't gotten us all that far. So, again, it comes down to the motherhood aspect of the issue. You know, "Boys, boys, why are we fighting here? We're all in this together. We all have the same goals. So why can't we come up with solutions?" Well I guess if that's true then we make a valid point that if we are really committed to the same goals, and we're all really on the same team, we should be able to come up with something that we can all live with. Logically that does make sense. So we took that at face value and so we said "OK" because at the time . . . the thinking . . . was we're a lot smarter than everybody else. . . . We can out negotiate people. We're tougher. We're smarter. We've got more discipline. If we get into a negotiating situation with employers and government on an equal basis we'll do OK on that. We can use our wits and our intelligence and our knowledge and all of the skills that we have and the power base that we have; we can be winners in that equation. (Ublansky interview)

As we will outline later, the decision by the labor movement to engage in bipartist modes of operating proved to be an important factor in the demobilization of the OHS movement in the 1990s. In that moment, however, the effect of a turn to more bipartism was a matter of intense debate within the labor movement. The issue came to a head when the ruling Liberal government of David Peterson, supportive of this change of direction and alive to the pressures emanating from "grassroots" OHS conflicts like de Havilland and McDonnell Douglas, introduced Bill 208 in January 1989 with all of its bipartist provisions. OHS activists spoke out publicly against it. However, with the leadership of the labor movement firmly committed to bipartism (Roberts 1989; De Carlo 1989; OPSEU 1989; "Backing" 1988–89), Bill 208 was passed and became law at the beginning of the 1990s. A new era in OHS regulation, with new forms of worker participation at its center, was seemingly in the offing.

Writing the Workers Out

The 1990s opened with the surprising electoral victory of the Ontario NDP. In 1985 the Liberal Party had ended the forty-three-year unbroken reign of the Progressive Conservatives when, after winning a small plurality of seats in the election, it entered into an accord with the NDP to form a coalition government. This always uneasy alliance came undone two years later when the Liberal Party won an overwhelming majority of the seats in a provincial election. Thus, it was surprising both that an election was called in 1990 (the Liberal government was not legally required to call an election until 1992) and that in the election the NDP emerged with its own majority government. Once over the shock, however, trade unions and other social movement groups in Ontario readied themselves to reap the benefits of having a sympathetic party holding the economic and political reigns of power.

Unfortunately for these individuals, groups, and social movement organizations, history was not about to unfold according to hopes and expectations. After forming the Ontario government for the first time in its history, the NDP ran up against two seemingly intractable forces: an economy sliding quickly into severe recession, and a business community determined to oppose any and all of the new government's social, economic, and political reform efforts. It was, thus, in this context that a social democratic government, already predisposed to governing from the perspective of "all Ontarians," moved slowly with its legislative agenda—taking, for example, nearly three years to enact labor law reforms, including limits on the use of replacement workers and the extension of collective bargaining rights to farm workers and domestics.[12]

This slow pace of initiative and change upset political and trade union allies alike. However, as many of our interviewees stated, the poor state of the provincial economy prompted them to refrain from openly expressing their frustrations and mounting a more sustained critique. This wait-and-see attitude changed drastically, however, when the government, in a self-proclaimed act of trying to tame the provincial deficit, imposed a "social contract" on the broader public sector that required workers to take days off without pay. In one fell swoop, the table was set for deeply divisive conflicts between the NDP and many of its important friends and allies within the labor movement.

The NDP's record on workers' compensation and OHS was a microcosm of its overall approach to governing. Because of its embrace of bipartite consensus politics, the government failed to act when employers resisted proposals to reduce exposure limits to hazardous substances or to require longer training periods for certification of worker and employer members of JHSC's. Just as it became obsessed with the politics of deficit reduction, it also accepted the need to reduce costs within the workers' compensation system because of an unfunded liability, leading it to partially de-index the pension benefits of permanently disabled workers. In exchange for this financial sacrifice, workers were given an equal number of seats as employers on the governing board.[13] Finally, the enforcement of OHS laws, including a reduction in inspections, orders issues, and prosecutions, fell to record lows (Tucker forthcoming).

Having alienated many of its supporters and effectively renouncing the possibility of a social democratic alternative, it was not surprising that the NDP was swept from government in 1995 by an unapologetically right-wing Progressive Conservative Party (PC). The new government immediately embarked on a political mission designed to dismantle the central pillars of the province's welfare state and to refashion labor and employment law in a manner that made it far more congenial to business interests. With respect to the latter, not only did it repeal the NDP's collective bargaining reforms, but the PC government introduced additional changes designed to make it more difficult for workers to unionize and obtain a collective agreement (Jain and Muthu 1996). In the area of workers' compensation, the PC government abolished a royal commission on workers' compensation that the NDP government had appointed late in its term, and ended bipartite governance at the WCB.[14] It also initiated a review of the workers' compensation system that ended in a complete rewriting of the law in 1997.[15]

A number of these changes to the workers' compensation system are important to discuss as they have had a profound impact on workers and unions and their orientation to OHS. Among these changes the early and safe return to work (ESRW) and labor market re-entry (LMR) plans are perhaps the most significant. Put forward as changes that would encourage—in a positive fashion—injured workers to return to work, and/or assist them in finding news jobs should they no longer be able to work at their previous occupations, the ESRW and LMR plans drew immediate and strong criticism from injured workers and trade unions. With regard to the ESRW, they charged that workers would be coerced—subtly and otherwise—to return to their jobs before they were completely healed and psychologically ready.

Moreover, they argued that the LMR scheme would result in injured workers being placed in jobs inferior in quality and income to those they had performed in the past. Subsequent studies of these programs support these concerns. For example, in an examination of the operation of the ESRW program in the small business sector, Joan Eakin and her colleagues (2003) underlined the pressures felt by injured workers to return to their jobs long before they felt capable of doing so. While the researchers are quick to point out that such pressures are related to the tight operating budgets characteristic of the small business sector, where employer-employee relations are routinely complicated by ties of family and friendship, the findings of their study nevertheless affirm the initial (and ongoing) fears of injured workers. These results are echoed by the union-based OHS officials interviewed for the present study. According to Tom McKay, for example, an assembler in a transnational plastics plant on the outskirts of Toronto, and an OHS activist with the Union of Needletrades, Industrial and Textile Employees (UNITE), the ESRW program has resulted in "people being put back to work before they are ready" (McKay interview).

Employer enthusiasm for ESRW is fueled by the expansion of experience rating within the workers' compensation system, which links premiums to employers' individual claims' costs. ESRW is only one dimension of employer claims' management; another that has immediate significance to OHS is not reporting injuries or misreporting them as no-lost-time claims when they ought to have been reported as lost-time claims (Thomason and Burton 2000). Not only does nonreporting benefit employers, but it also allows the government to make exaggerated claims about the efficacy of its OHS strategy. For workers and unions, however, these forms of claims' management are clearly problematic. In response to a question probing the main OHS concerns of his union, Cam Sherk was quick with the reply that his "major concern is the fact that workers are not reporting accidents that happen while they are at work and on the job" (Sherk interview). While this is not a new phenomenon, particularly for unorganized workers, it makes the job of union OHS activists more difficult still, as not only do they confront the opposition of employers, but increasingly they must also find ways to overcome the hesitancy of their injured members. Moreover, this reluctance of workers to report their injuries is in many instances being preceded by an unwillingness of workers to confront employers over hazardous working conditions. Claudine Salama, an OHS activist with the federal government employees' union, the Public Service Alliance of Canada (PSAC), compared the current fears of workers with previous examples of bold and decisive action.

> Today they're not very active and I think the reason why they're not very active is that they're very fearful to lose their jobs because of cutbacks. . . .

> There's a climate that exists in the labor movement, I think, and within our government, the department, and the workers today where they really feel powerless and they feel very angry. And they are also discouraged from asking for changes because of the competitiveness. And I'll give you an example. A few years ago, about 1987, there was a problem in a government building around health and safety with asbestos and at that time I was successful as a health and safety officer to get every single worker out of this building. To walk out because of the asbestos and we were out on the street. . . . We had the same situation last year . . . [in] another government building and despite all of my effort, the same kind of effort and more to get workers to walk out, they did not. They're working because they don't want to jeopardize their jobs. The thing is, the activism is not there any more the way it used to be. (Salama interview)

These changes, along with others (including shutting down the IDSP and the renaming the act that removed "worker" and "compensation" from its title), were patent examples of the neoliberal thrust of the PC government; and, there was more threatened. Indeed, at the same time that the government was proceeding with compensation "reforms," it was also preparing its OHS strategy. At the center of this set of initiatives was one component to which it was absolutely committed: the end of regulatory bipartism. Shortly after taking power, the Progressive Conservatives' government started dismantling the WHSA along with a labor-management advisory group that had been established under the Liberal government to develop hazardous substance regulations.[16] The government also announced budget cuts to the Ministry of Labour that called for a 20 percent reduction in the number of inspectors (Ontario Ministry of Labour 1995). It also launched a series of reviews beginning in 1995 that led to the release of two government discussion papers, *Review of the Occupational Health and Safety Act* (February 1997) and *Preventing Illness and Injury* (January 1998). These documents set out to refashion the principles that underlay any changes that had been made to the OHS Act after its implementation in 1979. As Eric Tucker has written elsewhere, the first document stated that the two main principles to be rearticulated were that the act should support and create self-reliant workplaces where employers and employees cooperate to create safe and healthy workplaces, and that the role of the Ministry of Labour was to ensure that employers with poor health and safety programs and performance comply with protective standards. As Nichols and Tucker (2000, 301) state, "The principle supporting this approach was laid out in the document under the heading 'Economic Growth,' where it stated that 'competitiveness and safety excellence are complementary and mutually reinforcing goals.' In short, what the documents were stating was that 'since safety pays, rational, informed and competent employers will voluntarily provide safety.'"

Not surprisingly, the issuing of these documents set off worried rumblings within the OHS community. Already alert and alarmed by the government's closing of the WHSA, the winding down of the Industrial Disease Standards Panel (IDSP), and the attempted termination of the Occupational Health Clinics for Ontario Workers (OHCOW), OHS activists both within and outside the labor movement carefully scrutinized the two documents. Indeed, many unions issued extensive critiques of the discussion papers, including the CAW, which published a point-by-point condemnation of the government initiative. In the cover letter to this document, Basil Hargrove (CAW 1997), the president of the CAW, wrote in angry tones that the government's proposals constituted a "wholesale attack on workers' health and safety rights."

It proposes eroding the worker's right to refuse work that is dangerous, eroding the right to know about workplace hazards, and eroding the right to participate in joint health and safety committees. As well, it strongly emphasizes deregulation by proposing vague, "performance" regulations that would allow employers to do pretty much what they want. And, finally, it proposes eroding the Ministry of Labour enforcement by, for example, lowering fines and decreasing inspections of less hazardous (read, unionized) workplaces.

Within their document, the CAW took specific aim at the government's attempt to further align "competitiveness" and health and safety. "'Competitiveness' is not a health and safety goal," it stated, "nor is 'prosperity' or the 'overall economy' or 'productive' workplaces or 'job creation' and 'economic growth.' All of these are economic goals and have no place in a discussion paper about health and safety. . . . By accepting [these] ideas in any way, we run the risk of seeing each health and safety improvement in a workplace subject to a cost-benefit analysis." Moreover, the CAW (1997, 13) took critical aim at government claims that proposed changes to the IRS would not take away or diminish the rights of workers.

The government claims that one of the purposes of the review is to establish and maintain the internal responsibility system. We disagree. The review proposes weakening the major parts of the internal responsibility system that provide for worker rights. It also proposes weakening the responsibility of employers to comply with the act by proposing to reduce fines for noncompliance. In the context of substantial government cutbacks to the Ministry of Labour, it means that the government is attempting to get "out of the government business," allowing business to make even more decisions in health and safety than at present.

The government claims that flexible legislation is a goal that would allow the "workplace parties" to exercise greater discretion on how standards would be met. This is an enormous change. It could be as broad as allowing a company to decide that in order to ensure that workers are exposed to levels of airborne contamination below the legal limit, workers must wear respirators instead of installing a ventilation system. The government is proposing to weaken the joint health and safety committees, which will give employers more power and workers less input. Ironically, these joint committees today are already so weak that they have no power to make a decision but instead must make recommendations to the same employer who already holds half the committee positions.

Notably, the government did not follow up its discussion papers with major legislative changes. Nor did it use the occasion of budget cuts to reduce the inspection force. Indeed, in 2000 it announced new, lower exposure limits for 202 hazardous substances and committed additional money for their enforcement. Moreover, inspection data obtained from the Ministry of Labour for this period show an *increase* in the number of inspections and orders issued from those years when the NDP was in power (Tucker forthcoming).

Does this signal that OHS was largely spared from the PC government's neoliberal assault on labor and employment law? According to the OHS activists and officials the government's data are highly suspect. On one hand they claim that inspections carried out under the PC government's reign were less thorough than in previous years. Moreover, they charge that the reported increase in the number of inspections stemmed from new inspection procedures whereby government inspectors targeted and blitzed all the workplaces in industrial parks in a single day or even one afternoon. The end result was that large numbers of workplaces were inspected, but not in a manner that was likely to discover hazards and violations, as OFL-OHS director Vern Edwards detailed in an interview:

> Well I think there has been a change. They made a conscious decision that they will increase statistics so I think there's this focus on increasing statistics in terms of visits and inspections and whatnot. But, from what I've been hearing from OPSEU and even from some inspectors, quietly because they won't say this openly because they'll be fired, is that the quality of inspections have changed. All the ministry wants is data. They want numbers. So, they'll do blitzes on say an industrial park or one of these industrial plazas where they can just go door to door to door. Boom. Boom. Boom. They don't do any sort of substantive assessment of the workplace and so that happens a lot. And they seem to do these drives, I think, when they haven't reached . . . I don't know what kind of quota they have. Sort of like, you know, the police. If they don't have enough tickets by the end of

> the month they've got to go out there and they're doing blitzes and they get their stats up. We know they have quotas.

The most serious legislative changes were made in 2001 when, essentially without warning, the government amended the OHS Act (Bill 57) to allow for Ministry of Labour inspectors to mediate work refusals over the telephone instead of having to physically visit the worksite.[17] Although on its face this change to the process of mediating the right to refuse could be seen as merely technical, this was not how OHS activists and union officials perceived it. Rather, they saw it as a method of putting more distance between workers and the government in this critical area of worker participation. It seemed to be a validation of the CAW worry that the government was getting out of the business of governing—a development that profoundly worried the OSH community as it seemed to tilt the balance of power even further in the direction of employers. As OHS activists already understood, and as various studies on the right to refuse have documented, the refusal process was already not working in the interests of workers. In contravention of the act, employers in Ontario had disciplined or otherwise punished their employees for utilizing their right to refuse unsafe work (Harcourt and Harcourt 2000; Walters 1991). Moreover, in an insightful and important study of a unionized auto parts assembly plant, Garry Gray (2002) has shown how difficult it is for workers to use this right. Indeed, as Gray's study and our interviewees inform us, workers feel pressured—from their employers *and their coworkers*—not to refuse unsafe work. In speaking about the right to refuse, Tom McKay related how it took courage for a worker to refuse what he or she perceived to be an unsafe job.

> I think that overall people really don't feel good about it. I think that even though they have been educated there is fear of reprisal. Some of that is probably justified. And, I think overall people don't want to go to that extreme to get something changed. So, I think it is uncomfortable. It takes them out of their comfort zone. (McKay interview)

With Bill 57, then, the government seemed to be signaling that it was distancing itself from the day-to-day operation of the IRS. In conjunction with its record on prosecutions, this suggests that the PC government had adopted the approach of leaving employers free to self-regulate within the framework of a loosely enforced statutory IRS, but that if a worker was killed or seriously injured because of noncompliance with the law, then employers should expect to be prosecuted and face significant fines.

Health and safety activists interviewed for this study uniformly agree that health and safety has suffered under the PC government, but they remain divided in their views on the merits of the IRS system. As we have already noted, many in the OHS activist community criticized the faulty assumptions and operation of the IRS since its inception in 1979. For these individuals, the past years under the PC government have done little to change these long-held views. Colin Lambert, long-time national director of OHS for the Canadian Union of Public Employees (CUPE), is among those who believe the IRS was "doomed from the start."

> I think it was doomed from the beginning because there was never any intention to enforce internal responsibility. It's never been enforced anywhere in this country. I don't know of one employer that's been taken to court, fined, put in jail or anything else because they discriminated against a worker who used their right to refuse. If we're saying the basic building blocks of the law is internal responsibility, which means the right to know, the right to participate and the right to refuse, and you don't do anything about those when they're violated, then you don't mean it to work. You absolutely don't; just the opposite, it's just a joke. If that's what's going to drive health and safety in the workplace then its like the speed limits on the road, right? You have to enforce it 'cause that drives safety on the roads. That's the main thing that drives safety on the roads. So if you're not going to enforce it ever. . . . As I say . . . no one's ever gone to jail for not giving workers the knowledge that they're entitled to. No one's gone to court for that. No one's gone to court for not having a health and safety committee and some have not had a health and safety committee for years and years and years, right? None have gone to court for having a health and safety committee then denying them the right to operate properly. None. Ever. Anywhere. So it was doomed from the beginning and it was deliberate. Had to be deliberate' cause if you've got this wonderful theory and you want it to work then you'll make sure that the driver of this theory is doing his job, right? (Lambert interview)

Gary Cwitco, the OHS director for the Canadian Energy and Paper Workers Union (CEP) during the 1980s and early 1990s, agrees: "The government was never prepared to provide the authority that goes along with the responsibility" (Cwitco interview). Claudine Salama's views are in line with Lambert's and Cwitco's: "The IRS does not work," she stated firmly, "because there is, right away, a problem with the power relationship."

> You can talk about an internal responsibility system, but if you don't give the workers equal power or a budget to deal with issues of occupational health and safety you are making the whole concept not likely to work.

> Unless workers have a say on how work is organized and how much money they're going to spend on ergonomic equipment, as an example, then what is the use of an internal responsibility system? How effective could it be? It's very deceiving you know; it deceives workers to believe that they are equal when it comes to making decisions on joint health committees. You know, when they are supposed to have equal power when in fact they don't. That makes it not workable. (Salama interview)

The other major strand of thought within the OHS activist community—that the IRS was, in fact, workable but was not being adequately enforced—also remains prevalent within the OHS activist community. Vern Edwards believes that the IRS can and has worked:

> So the internal responsibility system in the Ministry of Labour was simply, it sort of encouraged or coached employers to reach ethical compliance. Not just regulatory compliance, but ethical compliance, which sounds all warm and fuzzy but the reality is workers become abandoned by the government to the mercy of the employer. The internal responsibility system [is] not actually in the act. It's more of a Ministry of Labour philosophy and principle that they follow and it's sort of there in the actual things like joint health and safety committees and the role that they play. But it's not spelled out in the legislation itself. So without strong external enforcement, the internal responsibility system is really leaving workers to the mercy of employers. So unless you have people who are well trained, they're assertive and basically protected by a union, it's not going to be very effective. In nonunion sectors if workers raise health and safety issues they just get fired. That's it, end of story, bye. (Edwards interview)

In a somewhat different vein, another long-time OHS union activist decried the "anti-internal responsibility union's analysis" that centers on the "same concerns of mine: inertia, obstruction by employers, deregulation and lack of enforcement by governments, bureaucratization of health and safety within unions, a disempowering of workplace activists, etc., among other things." The IRS "has its faults," this activist conceded, "but it can be improved by emphasizing rights, roles and responsibilities, problem solving and decision making."

Stepping back from these two sets of views, the IRS experience in Ontario over this period can be summarized as follows. First, the effectiveness of what many (e.g., Bob DeMatteo) call the "eternal responsibility system" has "always depended on the activism of workers and unions" (see also Smith 2000). In short, as the experience of workers using the right to refuse unsafe work highlights, the IRS has been most effective in workplaces that are both unionized and where the union has itself emphasized OHS (Lewchuk et al. 1996; O'Grady 2000). Second, according to Andrew King, national director for OHS for the USWA, "what the whole internal responsibility system did provide was the opportunity for some activists, particularly those who had the support of their union, to take on certain kinds of projects and . . . use it as a lever." But there is, he argued, a critical "disjunction between [IRS] rights and the day-to-day concerns of the workers."

> Yes, well, I mean I think we'd better look a little bit further than do they know their rights. That's part of it. [But] the rights don't actually address a lot of the day-to-day concerns they have to deal with because the rights assume certain things about how health and safety problems emerge that aren't exactly accurate reflections of what workers experience. And we set up a system of representation that separates the representative from the workers and so there's a bunch of organizational problems here that overall undermine the system from working. But, having said that, there were worker's reps and workers who managed to break through in different places, make use of it, get support from different places and actually make some things happen. (King interview)

Conclusion

The results of the most recent provincial election held in the fall of 2003 were not surprising: the PC government was soundly defeated by a resurgent Liberal Party. In the election campaign the Liberals promised to reverse the rampant cost cutting in health and education, while stating that they intended to restore humane understanding in deliberations relating to other social policies and programs. No such promises were made, however, to the province's workers—except to say that economic measures designed to promote the economy would be beneficial to all. And, last, no mention was made of unions and their place in a refashioned big "L" liberalism.

Recently, however, the new minister of labour has established health and safety action groups in the construction, health care, and manufacturing sectors on which unions and employers are equally represented. In making the announcement regarding the establishment of the manufacturing sector action group, the minister of labour proclaimed that OHS needed to be improved both to strengthen the Ontario economy and to improve the quality of people's lives (Ontario ministry of labour 2004). At the same time, the PC-appointed CEO of the WSIB has resigned from his position amidst concern that he had spent WSIB monies for private luxuries. The government also initiated an audit of the WSIB's revenues and spending. In short, the winds of neoliberal change seem to have lost some force and the future of OHS and WSIB policies and programs seem open to challenge and change.

Yet if there is indeed space for agency within this shifting neoliberal paradigm, the content and the direction of this agency is a subject of great contest within the OHS community. The IRS remains the central point of debate precisely because it is unlikely to be replaced by any other form of regulation in the near and/or distant future—a position even the PC government was forced to accept. It is, thus, highly unlikely that the three health and safety action groups will put forward recommendations aimed either at eliminating or radically restructuring the IRS system.

Some parts of the broader OHS activist community are in motion. As a response to the three government action groups, the OFL has reactivated its OHS Committee. Initial meetings have focused on how the labor movement should respond. For Cam Sherk, a member of the OFL-OHS Committee and a member of the Manufacturing Health and Safety Action Group, the priorities are clear: There is a need for more inspectors and these inspectors must be given clear authority to do their jobs; members of Joint Health and Safety Committees require better training; and employers must be obligated to act on the recommendations of the joint committees.

These priorities are the same ones held by virtually every OHS activist interviewed for this study. Almost without exception—and, no matter which side of the IRS divide they stood on—OHS activists were firm about the necessity of strengthening external (i.e., government) enforcement of the act. Ironically some pointed to the prosecutions of employers under the PC government to support their contention that such actions are politically feasible. While a few (e.g., Walker) perceived these activities as consistent with a "law and order" mentality of the PC government, they also suggested that they serve as a precedent for present and future governments. There is also a consensus on the most pressing OHS issues: mental health, musculoskeletal injury, industrial disease, worker's compensation, and management health and safety systems (MHSS).

OHS activists pointed to lean (and mean) production systems creating situations of high stress for ever-increasing numbers of workers. Workers also face psychological difficulties arising out of increasing job insecurity and for women in particular there is the additional stress of the work-family relationship (Carr interview; Williams 2003). In addition to stress, work intensification has produced a steady increase in the number and severity of musculoskeletal injuries.[18] Indeed, strains and sprains have accounted for approximately 50 percent of all lost-time compensation claims in Ontario for the past decade. The sheer numbers of these claims point not only to the close relationship between these new forms of production and injury and chronic pain, but they undermine earlier (and ongoing) claims by industry and health officials that such injuries arise out of either the fertile imaginations of women workers or were restricted to a small number of jobs (Kome 1998; Mogensen 1996; Hopkins 1989).

Occupational disease was also identified as an issue that will not—unfortunately and tragically—go away. Directly after the PC government closed down the WHSA and pink-slipped the IDSP in the early years of its reign, union activists within the CAW and the CEP helped mobilize hundreds of workers and their families in Sarnia, Ontario, who were stricken with various illnesses and diseases associated with asbestos exposure. As government documents uncovered by the union activists revealed, workers at an auto parts plant, Holmes Foundry, had been for decades exposed to levels of asbestos that vastly exceeded the legal exposure standards (Mittlestaedt 2004; Brophy and Parent 1999). The government was pressured to fund the creation of an occupational health clinic in the Sarnia area to attend to these workers and their families. At the same time, the WSIB has been forced to address a veritable avalanche of claims from workers and their families from this area.[19] While these claims arise from "old" exposures, they serve to remind OHS activists that the effects of current exposures to hazardous substances and stressful working conditions often manifest themselves in the future. Thus there is a real concern that occupational disease, far from being relegated to the past, is appearing in new forms in a vast array of jobs and workplaces.

The fourth area of concern relates to the development and operation of OHS management systems. In principle, these systems aim to improve employer OHS performance by making health and safety more central to management processes.[20] Participants in this study, however, saw their promotion as a sign of the withdrawal of governments from direct regulation and the growing power and confidence of business. In their experience, these systems are driven by employers' desires to reduce workers' compensation premiums, leading firms to focus on claims management. As discussed earlier, this often results in the nonreporting of work injuries. In fact, Cam Sherk claimed in 2004 that this is the most vexing problem faced by his union.

Regardless of the specific area of concern, the overriding lesson to be learned from the recent history of OHS in Ontario is that victories were only won because there was a vibrant and militant grassroots health and safety movement. Paradoxically, to the extent that the successes of the OHS movement crystallized in bipartist institutions and forms of education and training, they reoriented OHS activists and trade union officials *away* from the rank and file and the union local toward paid union staff and a centralized OHS education/training model and delivery service. In these ways OHS was removed from the workplace—figuratively and literally—and plunked down in classrooms where, critics of this evolution charge, the political content of the courses has been replaced by an emphasis on the technical and scientific bases of health and safety.[21]

How, then, to regain the "momentum?"[22] For the overwhelming majority of the activists interviewed for this study OHS must become what one individual termed a "core issue" for trade unions. But, how to accomplish this goal with, as Tom McKay related in 2001, trade union activists "hav[ing] so much work there that it is hard to do the other things. . . . Just the . . . things that it takes to get [OHS] outside of the plant. Because you are swamped inside the plants." One suggestion offered by a number of OHS activists was for local unions to establish their own OHS committees within the workplaces they represent. Only workers would staff these committees—no management representatives would be permitted. Further, these committees would be required, like negotiating and political action committees, to submit monthly reports to local union meetings. In these ways, it is argued, OHS would become an integral component of union life. In tandem with these proposals is another that places the principal responsibility for formulating and delivering OHS education and training in union locals. According to D'Arcy Martin in 2001, taken together these proposals would serve both to redress the excesses of the OHS movement's institutionalization and help fuel its repoliticization.

At bottom, the repoliticization of the OHS movement is understood as being coterminous with the mobilization of the rank and file that, in turn, was unanimously perceived to be the main course on the "what is to be done" menu. But this prescription begs the further question of how what needs to be done is done. The formal politics of neoliberalism in Ontario may have been sent packing in the latest election, but this should not blind us to the fact that while the neoliberal train was in the station its agents managed to alter many economic, political, and institutional realities, as well as key sociological sensibilities relating to the importance of and responsibility for physical and psychological well-being. Self-regulation and the promotion of management OHS systems, for example, have the effect of placing the responsibility for accidents and injuries on the workers themselves. This is, in essence, little more than a new, albeit more sophisticated, version of blaming the victim. The battle cry of the 1970s and 1980s OHS movement was "Our health is not for sale." The challenge for present-day OHS activists is to reinvigorate that refusal by mobilizing workers around the OHS issues that are most important in the opening years of the twenty-first century and to reshape popular discourse so that workers' lives and health are given priority over the rapacious bottom line and commodifying dynamics of neoliberal, capitalist societies.

Notes

1. The phrase "occupational health and safety" is, in some respects, an inaccurate representation of the issues, events, and processes to be described and analyzed in this chapter. A better phrasing would be "workers" or "workplace" health and safety, thereby broadening the concept to encompass its critical social, economic, political, and environmental dimensions. However, as the movement evolved, it was termed and came to be generally known as the "occupational" health and safety movement. Hence, that is how it will be referenced here.

2. These have been described elsewhere as "third-wave" reforms, following the initial choice of market regulation in the mid-nineteenth century and direct state regulation in the late nineteenth and early twentieth centuries (Tucker 1995, 245–67).

3. This view is echoed, in a more theoretical fashion, by Robert Sass (1989, 168–69):

> In the case of the work environment, existing workers' rights "to know, participate and refuse" belong to individual employees rather than to workers in a work setting consistent with the social nature of production. The very language of "rights" afforded workers in law is permeated with possessive individualism and rests upon a claim based on liberal contractualism, which emerged with commercial law and private property. An individual worker, therefore, has only legal claim against an employer rather than an entitlement as a member of a moral community that prescribes standards of conduct and care. The liberal conception of worker rights in workplace health and safety is based upon a worker's claim against other persons. Rights are merely personal properties: it is "our" rights, "their" rights, "his" or "her" rights that support a worker's claim for compensation for hazardous and dangerous work, and injury.

4. This approach predated the IRS system. For the role of inspectors in Ontario, see Tucker (1988).

5. The interviews that are referenced in the following pages were carried out from 2000 to the present. At this point they include more than fifty interviews with OHS and injured-worker activists. While no claim is being made regarding their "representativesness," they include most of the individuals who are generally recognized to be among the key actors in the OHS arena over this entire period.

6. In the mid-1980s, the Ontario Public Service Employees Union, in defense of its government inspector members, published a scathing review of the operations of the health and safety branch of the Ministry of Labour charging officials within the upper echelons of the ministry with either stalling or sabotaging the work of the inspectors. This report led to the government appointing two corporate lawyers to conduct a review of the IRS. This report "cleared" the ministry of any wrongdoing and supported the IRS in philosophy and form (Ontario Government 1987).

7. The Occupational Health Movement (OHM) in Ontario in the 1970s and 1980s was composed of many different actors and organizations. The four most important were rank and file industrial workers, trade union officials, middle-class political radicals, and social democrats within the NDP. Each of these groups had somewhat different and, at times, conflicting political orientations to workplace health and safety. What ultimately led to their coalescence was that each could find meaning and promote activism in the phrase "our health is not for sale."

8. From the late 1970s to the mid-1980s, the PC government attempted to alter the workers' compensation system in ways geared to lowering employer assessments. In so doing, it confronted an organized and increasingly militant injured workers movement, which opposed these actions via countless individual and collective protests. Finally, in 1984, the government shelved the most controversial changes to the Workmen's Compensation Act, putting in their place a series of progressive changes, including one that provided for the establishment of an appeals tribunal that, for the first time in the history of workers' compensation in Ontario, allowed for an independent review of board decisions. The amendments also changed the name from "Workmen's" to "Workers'" Compensation Act. For a fuller examination of these developments, see McCombie (1984) and Storey (2004b).

9. The submission made by McDonald's Restaurants of Canada (1990, R-17), while stating its case more strongly than others, was nevertheless representative of other employer views on the need to have nonunion "directors appointed the new Workplace Health and Safety Agency. We recommend, since a large majority of the workforce in Ontario is nonunion, that a provision be made for the appointment of representatives who have no affiliation with the union."

10. In our interviews, there was virtual unanimity among OHS activists that the 1987–88 period was the time when Ontario employers began to mount their resistance.

11. Gary Newhouse, a workers' compensation lawyer and an activist in the movement in the late 1970s and 1980s, recalled being present at a meeting in the later 1980s with officials with the Ministry of Labour and leaders from the USWA, the CAW, and the OFL. In that meeting Newhouse heard these leaders tell the government that they could "control" their members but they could not control the injured workers and their leadership.

12. In his victory speech Bob Rae, the leader of the NDP, noted that even though his party did not win the majority of the popular vote, those who did not support the NDP could rest assured that it would govern the province with "all Ontarians" in mind.

13. Workers' Compensation and Occupational Health and Safety Amendment Act, 1994, S.O. 1994, c. 24. To head the Workers' Compensation Board, the NDP appointed Odoardo DiSanto, an ex-NDP member of the Ontario legislature and longtime advocate for injured workers.

14. Workers' Compensation and Occupational Health and Safety Amendment Act, 1995, S.O. 1995, c. 5.

15. Workplace Safety and Insurance Act, S.O. 1997, c. 16, Sch. A.

16. S.O. 1995, c. 5.

17. Government Efficiency Act, S.O. 2001, c. 9.

18. For comparable U.S. data, see Brenner, Fairris, and Ruser (2002); and Askenazy (2001).

19. One of the key activists in uncovering these tragic processes is Jim Brophy, an OHS activist in Ontario since the mid-1970s. In our 2002 interview Brophy commented that he did not believe his group could have won the support of the labor movement in this situation if Holmes Foundry were still operating.

20. For an excellent overview of its development and debates surrounding its efficacy, see Frick et al. (2000). For critical perspectives, see Nichols and Tucker (2000) and Øystein Saksvik and Quinlan (2003).

21. In our 2000 interview, Cwitco outlined his understanding of the these events in the following way:

> The employers became much more sophisticated. You know, instead of having safety engineers on staff they had industrial hygienists on staff. And as we became more sophisticated, they became more sophisticated, and, ultimately, it was where we lost control. The discussions became scientific rather than political around the resolution of health and safety problems. We debated the science rather than politics. We created that monster and we fell into the trap of allowing science to become the determining factor of whether or not something should be done. Employers always could get better science than we could or more ready access to it. There were lots of times that we would find the stuff and that's when we had a chance, but, as the energy that came out of the 1960s into the 1970s around activism and moved into the late 1970s and the 1980s, the people in the academic community who were doing a lot of that research were off doing other things and the kind of flow of information that was coming out of academic institutions supporting unions and workers stopped or slowed down at the very least. So that once we had created this battlefield of science we started losing a lot.

22. As part of a larger debate on the need to rediscover social movement unionism, there is discussion in the OHS activist community on the need for alliances—especially with the environmental movement. For a critical analyses of this topic, see Storey (forthcoming).

References

Askenazy, Philippe. 2001. "Innovative Work Practices and Occupational Injuries and Illnesses in the United States." *Economic and Industrial Democracy* 22: 485–516.

"Backing a Revised OHSA," 1988–1989. *At the Source* 9, no. 4 (Winter): 10.

Brenner, Mark D., David Fairris, and John Ruser. 2002. "'Flexible' Work Practices and Occupational Safety and Health: Exploring the Relationship between Cumulative Trauma Disorders and Workplace Transformation." University of Massachusetts Amherst, Political Economy Research Unit, Working Paper Series, number 30.

Brophy, Jim, and Mark Parent. 1999. "Documenting the Asbestos Story in Sarnia." *New Solutions* 9: 297–316.

Canadian Auto Workers Union (CAW). 1997. *Government Discussion Paper: Occupational Health and Safety Act*. Toronto.

Canadian Federation of Independent Business (CFIB). 1990. Submission on Bill 208 to Standing Committee on Resources Development, in Legislative Assembly of Ontario, January 15, R-5.

Canadian Manufacturers' Association of Canada, Ontario Division. 1990. Submission on Bill 208 to Standing Committee on Resources Development, in Legislative Assembly of Ontario, January 17, R-117.

De Carlo, Nick. 1989. "The Right to Refuse Bill 208." *Our Times* (May): 9–11.

Eakin, Joan, et al. 2003. "'Playing It Smart 'With Return To Work: Small Workplace Experience Under Ontario's Policy of Self-reliance and Early Return.'" *Policy and Practice in Health and Safety* 1, no. 2: 19–42.

Frick, Kaj, et al., eds. 2000. *Systematic Occupational Health and Safety Management*. Amsterdam: Pergamon.

Gray, Garry. 2002. "A Socio-Legal Ethnography of the Legal Right to Refuse Dangerous Work." *Studies in Law, Politics & Society* 24: 133–69.

Harcourt, Mark, and Sondra Harcourt. 2000. "When Can an Employee Refuse Unsafe Work and Expect to Be Protected from Discipline? Evidence From Canada." *Industrial and Labor Relations Review* 53: 684–703.

Hopkins, Andrew. 1989. "The Social Construction of Repetition Strain Injury." *Australian/New Zealand Journal of Sociology* 25: 239–59.

Jain, Harish C., and S. Muthu. 1996. "Ontario Labour Law Reforms: A Comparative Study of Bill 40 and Bill 7." *Canadian Labour & Employment Law Journal* 4: 311–40.

Kome, Penny. 1998. *Wounded Workers: The Politics of Musculoskeletal Injuries.* Toronto: University of Toronto Press.

Lewchuk, Wayne, et al. 1996. "The Effectiveness of Bill 70 and Joint Health and Safety Committees in Reducing Injuries in the Workplace: The Case of Ontario." *Canadian Public Policy,* 23: 225–43.

Livesey, Bruce. 1988. "Profits before People." *Business Journal* (April): 28–35.

Martel, Elie. 1995. "The Name of the Game Is Power: Labour's Struggle for Health and Safety Legislation." In *Hard Lessons: The Mine Mill Union in the Canadian Labour Movement,* ed. M. Steedman et al., 195–209. Toronto: Dundurn.

McCombie, Nick. 1984. "Justice For Injured Workers: A Community Responds to Government 'Reform.'" *Canadian Community Law Journal* 7: 136–73.

McDonald's Restaurants of Canada. 1990. Submission on Bill 208 to Standing Committee on Resources Development, in Legislative Assembly of Ontario, January 15, R-17.

Mittelstaedt, Martin. 2004. "Dying for a Living." *Globe and Mail* (March 13): F1–F5.

Mogensen, Vernon. 1996. *Office Politics: Computers, Labor and the Fight for Safety and Health.* New Brunswick, NJ: Rutgers University Press.

Nichols, Theo, and Eric Tucker. 2000. "OHS Management Systems in the UK and Ontario, Canada." In *Systematic Occupational Health and Safety Management,* ed. Kaj Frick et al., 285–309. Amsterdam: Pergamon.

O'Grady, John. 2000. "Joint Health and Safety Committees: Finding A Balance." In *Injury and the New World of Work,* ed. Terrence Sullivan, 162–97. Vancouver: University of British Columbia Press.

Ontario Government. 1987. *Report on the Administration of the Occupational Health and Safety Act.* 2 vols. Toronto: Ontario Ministry of Labour.

Ontario Ministry of Labour. 1995. "Notes on Ministry of Labour Expenditure Reduction Strategy," August 18.

———. 1997. *Review of the Occupational Health and Safety Act.* Toronto: Ontario Ministry of Labour.

———. 1998. *Preventing Illness and Injury.* Toronto: Ontario Ministry of Labour.

———. 2004. "Minister's Action Group Moving Quickly to Reduce Injuries to Manufacturing Workers," March 11. Available at: www.gov.on.ca/LAB/english/news/pdf/2004/04–29.pdf.

Ontario NDP. 1983. "Not Yet Healthy, Not Yet Safe." Ontario New Democratic Party Task Force on Occupational Health and Safety, Elie Martel, Chair.

———. 1986. "Still Not Healthy, Still Not Safe." Ontario New Democratic Party Task Force on Occupational Health and Safety, Elie Martel, Chair.

Ontario Public Service Employment Union (OPSEU). 1989. *Why we oppose Bill 208.* Pamphlet, May 29.

Øystein Saksvik, Per, and Michael Quinlan. 2003. "Regulating Systematic Occupational Health and Safety Management." *Relations Industrielles/Industrial Relations* 58: 81–107.

Roberts, Wayne. 1989. "Health and Safety Cast Adrift by Bill," *Now* (February 9–15): 13–18.

Sass, Robert. 1989. "The Implications of Work Organization for Occupational Health Policy: The Case of Canada." *International Journal of Health Services* 19: 157–73.

Smith, Doug. 2000. *Consulted to Death.* Winnipeg: Arbeiter Ring Press.

Storey, Robert. Forthcoming. "From the Environment to the Workplace . . . And Back Again? Occupational Health and Safety Activism in Ontario, 1970s–2000+." *Canadian Review of Sociology and Anthropology.*

———. 2004a. "Politics of the Body: OHS Activists and the Rise of the Occupational Health and Safety Movement in Ontario, 1960s–1980." Unpublished paper.

———. 2004b. "Our Only Power Was Moral: The Rise of the Injured Workers' Movement in Ontario, 1960–1985." Unpublished paper.

Thomason, Terry, and John Burton. 2000. "The Cost of Workers' Compensation in Ontario and British Columbia." In *Workers' Compensation: Foundations For Reform,* ed. Morley Gunderson and Douglas Hyatt, 261–98. Toronto: University of Toronto Press.

Tucker, Eric. Forthcoming. "Re-Mapping Worker Citizenship in Contemporary Occupational Health and Safety Regimes." *International Journal of Health Services.*

———. 1995. "And Defeat Goes On: An Assessment of Third-Wave Health and Safety Regulation." In *Corporate Crime,* ed. Frank Pearce and Laureen Snider, 245–67. Toronto: University of Toronto Press.

———. 1988. "Making the Workplace 'Safe' for Capitalism: The Enforcement of Factory Legislation in Nineteenth-Century Ontario." *Labour/Le Travail* 21 (Spring): 45–85.

Walker, Cathy. 1997. "NAFTA and Occupational Health: A Canadian Perspective." *Journal of Public Health Policy* 18: 325–33.

Walters, Vivienne. 1991. "State Mediation of Conflicts over Work Refusals: The Role of the Ontario Labour Relations Board." *International Journal of Health Services* 21: 717–29.

Williams, Cara. 2003. "Sources of Workplace Stress." *Perspectives on Labour and Income* 15, no. 3 (Autumn): 23–30.

Interviews

Anonymous (November 14, 2000)
Jim Brophy (June 24, 2002)
Joel Carr (December 19, 2000)
Gary Cwitco (December 12, 2000)
Nick DeCarlo (June 4, 2001)
Bob DeMatteo (January 12, 2002)
Vern Edwards (December 7, 2000)

Andrew King (December 6, 2000)
Colin Lambert (June 26, 2001)
Clarence Macpherson (January 12, 2001)
D'Arcy Martin (May 29, 2001)
Tom McKay (July 3, 2001)
Gary Newhouse (February 10, 2004)
Claudine Salama (January 8, 2001)
Cam Sherk (March 18, 2004)
Dan Ublansky (June 7, 2001)
Cathy Walker (December 1, 2000)

10

The Sinking of the Neoliberal P-36 Platform in Brazil

Carlos Eduardo Siqueira and *Nadia Haiama-Neurohr*

The forty-story oil-drilling-rig platform P-36 was the largest semi-submersible oil production platform in the world. It could operate in water depths as high as 4,462 feet, produce as much as 180 thousand barrels of crude oil per day (BPD) and 7.2 million cubic meters of compressed gas per day. Able to lodge 115 to 120 people, it was located in the Roncador oil field of the Campos Basin, state of Rio de Janeiro, Brazil, about 80 miles from the coast and 120 miles northeast of the city of Rio de Janeiro. The P-36 platform produced about 5 to 6 percent of the total oil production of Petrobrás, a state-owned oil enterprise—formerly a state monopoly—responsible for all oil production in Brazil. [1]

The P-36 platform started production in May 2000 and was producing about 64 thousand BPD as well as 1.3 million cubic meters of gas per day before the accident. The enormous platform sank in the Atlantic Ocean on March 20, 2001, after a couple of explosions on March 15 that killed eleven emergency responders who tried to extinguish fires in the platform. Most of the 175 workers who were on the platform at the time of the first explosion were evacuated. Emergency response crews in boats and helicopters were deployed to the scene of the accident to control a potential oil spill. The rig was surrounded by floating devices to prevent oil from spreading.

The slow sinking of the P-36 platform over a five-day period was well documented by the media. Brazilians and others throughout the world could watch step by step the unsuccessful measures taken by company personnel to prevent the platform from sinking. But what most will forever remember are the moving images broadcast by Brazilian television networks of a worker crying as the P-36 platform slipped beneath the waves of the ocean.

As soon as the platform sank, a national debate developed regarding the following questions: Why did the platform sink? Who was to blame for such a catastrophic disaster? Could the platform have been salvaged after the first explosion? Why did eleven workers die?

This chapter addresses these questions that were part of the national political and technical debate that was based on documents and reports produced by the Rio de Janeiro Oil Workers Union (Sindipetro), the United Federation of Oil Workers (FUP), the Association of Engineers of Petrobrás (AEPET), the Brazilian National Oil Agency (ANP), the Accident Investigation Committee of Petrobrás, and the Brazilian and international press, and on a few articles on offshore accidents published in Brazilian and American journals.

After a brief history of the P-36 oil-drilling-rig platform and the recent environmental and occupational incident records of Petrobrás, we provide a summary of the accident investigation they conducted. We argue that the root causes for the sinking of the P-36 platform lie in the neoliberal policies implemented by the Petrobrás management to restructure the company to reduce operation costs and increase production, following the same model that was adopted in the oil and petrochemical industry a decade earlier in Europe and the United States (Siqueira 2003). We conclude with a discussion of the management policies adopted in highly hazardous industries and facilities, because they may create serious public health consequences for workers, the environment, and the public. In addition, we briefly lay out the recent changes in Petrobrás after the election of President Luis Inácio Lula da Silva in November 2002.

A Brief History of the P-36 Platform

In 1996–97, a company called Marítima Petróleo Engenharia (Maritime Oil Engineering) was awarded contracts for the construction of six platforms worth $2.5 billion by the Brazilian oil company Petrobrás, without an open and competitive bidding process. Marítima is a small company with no previous experience building large oil platforms and whose Bolivian owner, German Efromovich, is a friend of Joel Rennó, a former president and a financial director of Petrobrás. In 1997, Efromovich brokered the lease of the P-36 platform, originally built in 1994, from the Italian Fincantieri shipyard for $354 million. He hired the Canadian shipyard Davie Industries to refit the original oil drilling "Spirit of Columbus" platform to make it an oil production platform, and the firm Noble Denton, also from Canada, for project design.

All of the drilling equipment and most production equipment had to be removed, and the operational capabilities in deep waters had to be increased from 100 to 500 meters of water depth up to 1,360 meters; 5.3 thousand tons of equipment and 3,000 tons of steel were also added to it. Months later the Davie shipyard went bankrupt. As a result, Petrobrás had to spend $200 million more to conclude the alteration of the P-36 platform in the Mauá shipyard in Rio de Janeiro. In October 1999 the more than $500 million platform left Quebec and headed aboard the Norwegian ship *Might Servant I* toward Rio where it arrived by February 2000 (Amaral 2001; Sindipetro 2004).

The P-36 platform had many technological innovations that led the Offshore Technology Conference (OTC) to award Petrobrás a Distinguish Achievement Award in 2001 "For outstanding advancements to deepwater technology and economics in the development of the Roncador Field; a timeline of 27 months from discovery to first oil production in a water depth of more than 1800 meters; made possible by the use of a dynamically positioned early production system, and a dedicated production system using steel catenary exporting risers, taut-leg polyester mooring, and subset production hardware" (OTC 2004). The P-36 platform also won the International Organization for Standardization (ISO) 14,000 and the British Standard 8,800 certifications for environmental and occupational health and safety management (Oficina de Informações 2001b).[2]

Thus, the history of the P-36 platform includes from the beginning all the signs of the global market-based investments characteristic of oil production: it was built in Italy and adapted in Canada, with parts shipped in a Norwegian ship and a Russian airplane, to be operated and maintained in Brazil by a mix of direct-hire and contract employees. In addition, it followed global voluntary corporate environmental and occupational safety and health standards.

Neoliberalism in Brazil and Petrobrás

While there are numerous definitions and approaches to the term "neoliberalism" (Ahumada 1996; Chomsky 1999; Tabb 2002; Brecher, Costello, and Smith 2000; Coburn 2000; Fiori 1998), we chose the following definition:

> As opposed to political liberalism—a term commonly used in the United States to define the philosophy of those who favor government intervention for the social good—the term neoliberalism, which is regularly employed in much of the world, should be understood to refer to a set of economic policies rooted in the old free-market economic liberalism epitomized by Adam Smith's *Wealth of Nations*. Today's neoliberalism can be defined as includ-

> ing the following components: the primacy of the market, the reduction in public expenditures for social services, the reduction of government regulation, and privatization of state-owned enterprises. (Karliner 1997, 2n)

The famous "Washington Consensus"—an expression created by John Williamson of the Washington, D.C.–based Institute of International Economics, to name the structural reforms proposed by the international finance community—became the guide and synthesis of economic reform in Latin America. Brazilian administrations fully implemented these economic and social policies in the 1990s together with the ideology of neoliberalism, a decade later than the United States and the United Kingdom, which adopted these policies in the early 1980s during the first Reagan and Thatcher administrations. The Washington Consensus aimed at stabilizing Latin American economies, reducing the size of the state, deregulating markets, liberalizing trade and finance, and opening national economies to trade and finance (Franco 1997).

Under the guidance of neoliberal policies, three successive Brazilian administrations successfully privatized and denationalized strategic state enterprises, such as large state-owned steel mills, public banks, and telecommunications and energy utilities. The last two Cardoso administrations (1994–2002) claimed to use money from the privatizations to raise needed revenues to shore up Brazilian foreign reserves and modernize the Brazilian economy, among other projects (Rocha 2002).

In 1995, a constitutional amendment broke the oil monopoly law. Law 9478 of 1997 (Petroleum Law) opened many oil exploration blocks to foreign investment and allowed the division of Petrobrás into subsidiaries. Multinational oil companies gained the right to own and export oil after payment of royalties. In 1998 the National Oil and Gas Agency (ANP) was created to regulate the Brazilian oil industry. The ANP has since carried out five licensing rounds of competitive bidding for exploration blocks located throughout Brazil. An ANP press release announcing the first Brazilian oil and gas licensing round states that Law 9478, along with the constitutional amendment,

> provided the authorization and roadmap for sweeping changes within the Brazilian energy sector through a comprehensive program of market-oriented reforms. These important initiatives comprised the removal of subsidies, import controls and price controls on crude oil, refined products and natural gas and ended the 45-year-old operating monopoly of Petrobrás.[3]

The Reichstul/Gros neoliberal administration of Petrobrás changed the structure of the company by creating forty decentralized business units and promoting competition among them to reduce costs and maximize production and profits. Reichstul advocated reengineering Petrobrás to reduce excessive hierarchical levels that allegedly created inefficient management. He also argued that Petrobrás should become an integrated energy company with subsidiaries and an international strategy to compete in the world oil market. He hired Arthur D. Little, a global management consulting firm, without competitive bidding, to develop the strategic planning for the company.

Fernando Siqueira, the former president of AEPET, stated that this planning "turned Petrobrás into a financial instead of an oil company." According to Siqueira, research and development teams were dismantled and the company has since bought "turnkey" technologies, which are technologies that have been built, installed, or supplied by the manufacturer complete and ready to operate without any input from professional experts. Wages went down for most employees while managers received bonuses and wage increases up to 100 percent, making the latter quite acquiescent to company policies that encouraged internal competition and downsizing (Oficina de Informações 2001a; Petróleo e AEPET 2001).

The Environmental and Occupational Health Record of Petrobrás

Before examining the specific environmental and occupational record of Petrobrás a brief review of occupational health and safety laws and regulations that apply to the oil industry is warranted. The legal framework for occupational health and safety regulations in Brazil is a complex web of federal, state, and local laws, standards, and ordinances. In spite of a formal corporatist tripartite regulatory system that includes labor unions, government agencies, and employer representatives, most regulations in the area derive from the activity of the Ministry of Labor, pursuant to Article 155 of Law 6514.

The majority of Brazilian health and safety regulations were written into chapter V of the Consolidated Labor Laws (CLT) of 1943—revised in 1977 by Law 6514—and the federal Decree (*portaria* in Portuguese) 3214 of 1978. This very important law approved twenty-eight *normas regulamentadoras* (NRs), or standards, to regulate workplace health and safety, forming the body of the Brazilian regulations, which also include executive orders dealing with a variety of International Labor Organization (ILO) health and safety conventions officially adopted, and a number of amendments to the NRs.

Law 6514 of 1977 changed chapter V of the CLT and established the general duties of government agencies, employers, and employees under the law. Titled "Safety and Occupational Medicine," it is divided into twenty-six sections, of which only two focus on the duties of employees and employers. Most of the sections have language about the powers of the Ministry of Labor to establish specific regulations in areas such as inspection of workplaces, personal protective equipment, the creation of company-level specialized services on safety and occupational medicine and preventive measures in occupational medicine, and penalty fees for violations of the regulations.

Frumkin and Câmara (1991) claimed that Brazil could be considered a good example of the discrepancy between reasonably good occupational health and safety laws in the books and loose enforcement in reality. Although many commentators in the United States also criticized the weak enforcement of regulations, especially during the Reagan and George H.W. Bush administrations, it appears that the overall level of noncompliance with regulations in Brazil is clearly higher. On the other hand, the Brazilian experience indicates that the presence of active unions in large workplaces increased compliance and enforcement of health and safety regulations in the 1980s. The 1990s was a decade of neoliberal deregulation in Brazil, when the Brazilian government reduced the enforcement of the NRs and environmental regulations (Siqueira 1998).

A short retrospective analysis of the major environmental and occupational accidents involving Petrobrás within the past decade reveals a history of negligence toward human health and safety and the environment. Environmental incidents reported by the international press show that major oil and fuel spills occurred in Brazilian waters in the past decade (Table 10.1). Environmentalists around the world were shocked when a tanker spilled 346,600 gallons of oil into the scenic Guanabara Bay in Rio de Janeiro. Six months later, the most disastrous environmental accident in the past twenty-five years occurred. One million gallons of crude oil from a broken pipeline were dumped into a major Brazilian river (the Iguaçú River) in southern Brazil (*World Environment News* 2001). This large oil spill happened in a refinery in the state of Paraná and was due to lack of adequate warning devices, such as modern oil pressure detectors and outdated equipment—the broken pipeline was over twenty years old (*BBC News* 2001).[4]

Moreover, numerous "small leaks" of oil and fuel have happened without public knowledge, because Petrobrás has considered them minor accidents without serious adverse consequences to the environment or human health. For instance, in June 2001, an oil spill contaminated an ecological reserve just outside of Rio de Janeiro for days.

According to the Sindipetro and the FUP, 102 workers have died between 1998 and 2001 at Petrobrás facilities around the country. Seventy-five of

Table 10.1

Environmental Accidents at Petrobrás Facilities, 1991–2001

Date	Location	Incident
September 1991	Coast of Rio de Janeiro State	Tanker dumped up to 150,000 barrels of oil
April 1992	Madre de Deus Refinery, Bahia State	12,800 gallons of oil dumped into the bay
August 1998	Sao Paulo State	Tanker dumped 4,000 gallons of oil
April 1999	Coast of Sao Paulo State	A faulty pipeline spewed 266 gallons of oil
August 1999	Coast of Bahia State	A broken pipeline spilled 13,300 gallons of oil
January 2000	Guanabara Bay, Rio de Janeiro State	A tanker spilled 346,600 gallons of oil
July 2000	Iguaçú River, Paraná State	A broken pipeline spilled 1 million gallons of crude oil
February 2001	Paraná State	A spill of 13,300 gallons of diesel fuel
March 2001	Campos Basin, Platform 36, Rio de Janeiro State	The world's biggest offshore oil rig exploded and sank

Source: Based on reports from *World Environment News* (2001), and *BBC News,* (2001).

those, approximately 74 percent, were contract workers (Table 10.2). For example, in December of 1998 three workers died during a fire at the Gabriel Passos Refinery in the state of Minas Gerais. Two more workers also died in January 2001 because of a fire at the offshore natural gas platform P-37 in the Campos Basin. The last in the series of fatal accidents occurred on March 2001 when eleven workers died after two explosions on the offshore P-36 platform (FUP 2001; *World Environmental News* 2001).

As a response to the large 2000 oil spills Petrobrás started the Program for Excellence in Environmental and Operational Safety Management (PEGASO), which aimed at: automated monitoring of all pipelines, reduction in hazardous waste generation and disposal, creation of Spill Response Centers, introduction of new technologies, and certification of its units by ISO 14,000 and BS 8,800 environmental management systems (Petrobrás 2004a).

Table 10.2

Fatal Accidents Among Workers at Petrobrás Facilities, 1998–2001

	1998	1999	2000	2001
Total victims	32	28	16	26
Percentage of contract workers	69	96	75	54

Source: Based on data collected by the Sindicato dos Petroleiros and FUP, October 2001.

Why Did the P-36 Platform Sink? Déjà Vu All Over Again?

After the sinking of the P-36 platform, a big national debate developed about the causes and reasons for the accident. The AEPET, oil workers' unions, professional organizations, the Petrobrás management, and wives of employees killed in the accident, among others, offered information to support their views on the immediate and root causes for the accident. The former president of AEPET noted that Petrobrás cut its budget for training by 50 percent (*Hora do Povo* 2001). FUP added that Petrobrás had recently downsized its workforce to nearly half (34,000) of what it was in 1989 when Petrobrás had about 62,000 employees (FUP 2001). According to Sindipetro there are 7,000 Petrobrás employees and 30,000 contract workers in the Campos Basin area; 130 of the 175 workers (or 70 percent of the P-36 workforce) on board the P-36 when the accident took place were contract workers.

Sindipetro and AEPET also denounced the worsened working conditions of the 1990s, during which permanent employees had to work very long hours and a two-tiered workforce emerged. The expansion of contracting out resulted in high turnover due to short-term contracts that did not offer adequate training, benefits, medical care, and social security. The two organizations considered the substitution of Petrobrás's untrained workers in operation, maintenance, and inspection services for permanent employees as the main cause for the series of occupational accidents that culminated with the sinking of the P-36 platform. Furthermore, both organizations blamed the Reichstul administration's neoliberal policies for the substantially increased number of fatalities and spills (Amaral 2001; Oficina de Informações 2001c; AEPET 2001b).

Woolfson and Beck (2000) performed a thorough analysis of the British regulatory system for the offshore oil industry operation after the world's worst offshore oil accident at Occidental's Piper Alpha oil platform in 1988. They pointed out that "the dramatic increase in incidents post–1985/86 can be directly attributed to the intensification of the labor process; and even more so, that there is an explicit correlation between the level of labor utilization and the levels of incidents" (44). They also argued that the Piper Alpha disaster in the North Sea was not "purely circumstantial," but that "it was the inevitable outcome" of intrinsically flawed regulations and an accompanying regime of labor relations. Their historical analysis of the structure of industrial labor relations in the offshore industry in the United Kingdom suggests that this structure allowed the British oil industry to capture and control the enforcement of health and safety regulations and to weaken labor unions.

After conducting a qualitative investigation of the causes of workplace accidents in the Campos Oil Basin, Freitas et al. (2001) concluded:

> The present panorama of work and safety conditions in the oilrigs reveals a situation of serious degradation, involving not only the potential for an increase in the frequency of accidents for workers, especially contract workers, but also the gravity of accidents, which may result in an accident with multiple fatalities. (2001, 127)

Those were almost prophetic words, based on evidence that indicated the less than remote probability of catastrophic accidents in Petrobrás's oil rigs in the area. A few months after the publication of the article the P-36 platform sank.

The U.S. Experience with Catastrophic Incidents in the Petrochemical Industry

In 1989 a Phillips 66 petrochemical plant in Texas exploded, resulting in twenty-three deaths and 232 people injured. In 1990, another incident, in an ARCO refinery, resulted in seventeen deaths (*Background* 1996, 1). These and several other catastrophic incidents involving highly hazardous chemicals (as defined by the U.S. Environmental Protection Agency, EPA) have drawn national attention to chemical catastrophes in the United States. The former Oil, Chemical, and Atomic Workers International Union (OCAW), now the Paper, Allied Health, Chemical, and Energy International Union (PACE), listed over twenty-two fires or explosions with fatalities in plants represented by the union between 1984 and 1991. The union called it the "body-count"-and-"disasters-on-the-rise" period. A study conducted by insurance industry consultants revealed not only that the number of large property disasters increased from twelve in the 1960s to fifty in the 1980s, but also that the costs involved skyrocketed to \$3.3 billion.[5]

After the Phillips 66 explosion and fire, the U.S. Occupational Safety and Health Administration (OSHA) commissioned a study by the John Gray

Institute of Lamar University, Texas, to examine the health and safety issues relating to the use of contract labor (nonpermanent workers) in the U.S. petrochemical industry. The investigation of the causes of the Phillips 66 incident disclosed serious problems regarding the training of contractor employees on chemical processes. Some of the recommendations of the study generated a heated debate between industry and labor regarding the safety and health consequences of the increasing use of contract labor or temporary workers in the industry. By the end of the 1980s, business and labor had also taken some measures to protect the public and workers from catastrophic chemical releases by creating environmental health and safety programs such as the Responsible Care Program (John Gray Institute 1991).[6]

On July 17, 1990, OSHA published the proposed Process Safety Management (PSM) standard containing requirements for the management of hazards associated with processes using highly hazardous chemicals, in order to assure that workers have a safe and healthful workplace. The OSHA proposed standard established a comprehensive management program and took a holistic approach that integrated technologies, procedures, and management practices (*Background* 1996, 2).

The former OCAW health and safety staff also developed their own comprehensive methodology to prevent and investigate catastrophic accidents, called "the systems of safety" method. PACE defines systems of safety as "the use of special management programs which actively seek to identify and control hazards (a proactive system). This begins in the conceptual (planning) phase of a project and continues throughout the entire process."

PACE divides the systems of safety into six main systems that have to be cared for to improve safety conditions in the workplace: the mechanical integrity system, the procedures and training system, the warning devices system, the process and equipment design system, the mitigation devices system, and the human factors system. Next, the chapter analyzes the most likely events that led to the sinking of the P-36 platform using the "systems of safety" incident investigation framework (PACE 1999).

The Petrobrás Accident Investigation Report

According to the report titled "Inquiry Commission P-36 Accident," the most probable sequence of events that led to the explosions and the ultimate sinking of the P-36 platform is:

> Excessive pressure in the Starboard Emergency Drain Tank (EDT) due to a mixture of water, oil and gas, which caused a mechanical rupture thus

> leaking the EDT fluids into the 4th level area of the column. The rupture of the Emergency Drain Tank caused damage to various items of equipment and installations in the column, principally the rupture of the sea water service pipe, thus initiating the flooding of this compartment, and released sufficient gas to fill the entire void space on the 4th level as well as other areas of the platform. After 17 minutes, dispersed gas—in contact with an ignition source—caught fire, causing a major explosion in the area where the firefighting crew was located and also resulting in serious physical damage to the platform. After unsuccessful attempts to stabilize the unit, the platform's increasing inclination—reflecting continuous flooding—resulted in the chain lockers and the vent tubes of the buoyancy tanks reaching sea water level causing progressive flooding, culminating in the loss of the platform. (Final Report 2001)

Another report, issued by the Brazilian Agencia National, "Análise do Acidente com a Plataforma P-36" (2001), shows a series of noncompliance issues that led to the P-36 platform disaster. Below we classify the noncompliance issues mentioned in the report using the systems of safety methodology.

The mechanical integrity system failures include systematic errors in the manual volumetric probing; malfunction of the EDT level indicators; mechanical failure or incomplete locking of the EDT entry valve; removal of the EDT evacuate bomb for repair the day before the accident; failure of the activators to close the sealed ventilation dampers; and absence of two seawater pumps, out of work for repair. The procedures and training system failures include deficient communication and coordination systems between the emergency response team and the command of the platform, and deficient coordination and training among the personnel in controlling the stability of the platform in an emergency situation.

The warning system failures include failure to set up portable gas detectors with sound and visual alarms for continuous monitoring. The design system failures include the requirement of the manifold production valve to perform its controlled opening with a secret code only accessible to the platform's coordinator; the vulnerability in the link between the EDT and the production manifold; the inadequate classification of the surrounding area of the EDT as a potential risk zone; and the use of a fail-set system that disabled the operator from modifiying its state because the system did not have an alternative system to skirt the imposed restriction. The mitigation system failures comprise the lack of gas detectors, vent gas scrubbers, and explosion-proof equipment in the tanker compartment of the third and forth levels. The human factors system failures consist of the decision to empty the EDT via the production manifold despite the operational instructions written in the Operating Process Manual; the emptying of the EDT without the supervision of the platform's coordinator or the production supervisor; the opening of the access ellipses to the ballast tank and the contiguous stability box for a longer-than-necessary time to perform the inspection and repair; and the inefficient actions to contain the damaged column flooding and to drain or perform water movement among nondamaged columns.

Fernando Siqueira criticized this report for not including the lack of maintenance and poor project design as important causes for the P-36 platform accident. He argued that because the P-36 platform was built overseas it started its operation too soon, without performing all the preoperation tests necessary to assess and prevent possible design errors. Thus, operators had only four months instead of the routine six to become familiar with equipment operation manuals. Several wives of operators who had been in Canada while the platform was under construction acknowledged that their husbands often complained about their lack of access to all areas of the platform. They also told investigation officials that only staff of Marítima, mostly English and Canadians, had access to certain restricted areas (Oficina de Informações 2001c).

Discussion

The evidence collected by the unions that represent oil workers in Brazil strongly suggests that there had been an increase in the numbers of deaths and injuries among Petrobrás direct-hire employees and contract workers during the 1990s. Above and beyond the specific chain of events that led to each and every accident or incident, the Brazilian oil worker unions claimed that there appears to be a root cause common to most if not all of those accidents: the restructuring of production and human resource management. The heavy losses in human life in the decade suggest that if one wants to solve the many failures diagnosed by the P-36 platform accident investigation committee, one has to move upstream—toward the root causes—in the chain of events that caused the explosion of the platform. To do otherwise (i.e., focusing on the proximate causes) is to delay or block once more the implementation of numerous operational and managerial procedures urgently required to prevent other catastrophic events.

Top-level management of the richest Brazilian company decided to manage the P-36 platform following the "rules" of the unregulated global market. A significant trend of this global market is to cut the size of the skilled workforce and replace it with less-skilled workers to increase profits and labor flexibility. In addition, management strongly demanded increased oil production to generate more revenue for the Brazilian treasury to compensate for growing public budget deficits (Sindipetro 2001).

This modus operandi is typical of what could be called the neoliberal management of a highly hazardous industry, that is, management policies that are driven by the political and economic requirements of neoliberal policies implemented by the Brazilian government. This "lean and mean" style of managing oil production may have increased short-term profits for the company, but it also intensified work and led to the hiring of poorly trained contract workers and lax maintenance of equipment. The combination of those three structural factors—intensification of work, hiring of poorly trained contractors, and lax maintenance—together with other ephemeral ones in the right time and in the right place could not but trigger dangerously unsafe conditions out of which catastrophic "accidents" happen (Druck 1999; Siqueira 2003).

This "recipe for disaster" lesson had already been learned by the offshore oil industry and unions after the Piper Alpha explosion in the North Sea in the 1980s. Woolfson and Beck (2000) note that in the latter case "the drive to cost-cutting and its effects on safety represent a recurrent cyclical phenomenon in the offshore oil industry" (44).

While it is true that Brazilian unions have waged similar specific battles for improving offshore working conditions and changes in management's policies, the main difference this time was that the sinking of the P-36 platform happened in a national context of growing popular dissatisfaction with the menu of neoliberal policies implemented by the Cardoso administration in the 1990s. The sinking of the P-36 platform was perceived by growing segments of the Brazilian population as another clear indication of the sinking of this neoliberal, "free-market" driven political, economic, and ideological platform that became hegemonic in the country during the 1990s (Petras and Veltmeyer 1999; Fiori 1997). Therefore, the sinking of the P-36 platform has become a good metaphor for the sinking of the larger neoliberal platform adopted by the Brazilian government and its elites.

The election of the Workers Party candidate Luis Inácio Lula da Silva in November 2002, who ran in an antineoliberal platform that proposed the reversal of the privatization and denationalization of state companies, started to change some important policies in Petrobrás. Under direction of Lula, the new Petrobrás administration has required that all new oil drill platforms (P-51, P-52, P-53) be built in Brazil and have a minimum of 65 percent of component equipment made in Brazil to create heavy industry jobs in Rio de Janeiro (*Jornal do Brasil* 2003a). The company has also increased the public opening of new positions for permanent employees in 2003 and 2004. Petrobrás plans to hire over 3,500 employees in 2004 to reduce the use of contract labor (*O Globo* 2003; *Jornal do Brasil* 2003b). By 2004 the PEGASO program has fifty-seven units of Petrobrás certified by either ISO or BS standards. Official company data indicate that the volume of spills decreased from 5,983 cubic meters in 2000 to 150 cubic meters in 2003 and that $2 billion will be spent in the program by 2007 (Petrobrás 2004b).

It is still too early to assess the breadth and depth of the changes implemented by Petrobrás in its health, safety, and environmental policies and practices. It took a few major spills followed by a major catastrophic incident for management to prioritize this area over the past four years. While the sinking of the P-36 platform may arguably have represented the end of the dominance of neoliberal ideas in Brazil, much needs to be done to move the country toward an alternative economic, political, and ideological path. There is much ongoing debate over how much change actually happened after the inauguration of the new Lula administration. Siqueira argues that many neoliberal managers remain in key positions and committees, which makes it very difficult for the new Petrobrás administration appointed by President Lula to reverse the company's dismantling (Siqueira 2003). In fact, many Brazilian political observers consider that this is the major dilemma of the Lula administration: to change the neoliberal economic policies followed by the two previous Cardoso administrations (Sader 2003; IHU On Line 2003).

The struggle to rebuild Petrobrás and shift its policies toward a different direction continues (AEPET 2004). Yet the recent changes in Petrobrás's human resources and investment policies bode well for future improvements in workplace health and safety conditions. Monitoring to what extent these shifts will become consolidated should be the subject of future research on the "reregulation" and "reconstruction" of state-owned companies in Brazil and South America.

Notes

1. Accessed November 25, 2002: www.petrobras.com.br/minisite/ingles/deep/units.htm. After the sinking of the P-36 platform the information is no longer available at this Web address, but there is information about the Roncador field at ww2.petrobras.com.br/ingles/index.asp.

2. The ISO is an independent federation of 100 national standards bodies established in 1946. The standards developed by the ISO comprise mainly product and process specifications. The national standards bodies are composed mainly of private interests, with, in some cases, government participation. The ISO 14,000 series certifications are given by ISO to companies that demonstrate compliance with its environmental management systems guidelines. British Standards is the national standards body of the United Kingdom. The British Standard (BS) 8,800 applies to occupational health and safety systems.

3. The same press release also defines the ANP "as [the agency] which has the

responsibility for the rational exploitation of the nation's petroleum resources and maintaining a fertile and responsive business climate which protects and balances the interests of both the private and public sectors." Available at: www.anp.gov.br/brasil-rounds/round1/HTML/Press_en.htm. Accessed February 22, 2004.

4. The P-36 had 316,000 gallons of diesel fuel stored aboard and 79,000 gallons of oil in the pipeline. See *World Environment News* (2001).

5. "Oil, Chemical, and Atomic Workers International Union, Process Safety Management Manual." This training manual is used to train PACE union members.

6. *Background* (1996, 13). This program originated in Canada in 1985 as a voluntary initiative and has expanded to include forty national chemical manufacturers' associations whose members account for 86 percent of world chemical output. Though the specifics of "Responsible Care" programs vary from country to country, they establish principles for manufacturers and distributors to continuously improve performance in all aspects of chemical safety, from the establishment of guidelines to the adoption of common logos and verification procedures. Adherence to Responsible Care principles is now a precondition for a firm's membership in the trade association in many countries.

References

Association of Engineers of Petrobrás (AEPET). 2001a. "Boletim 212. Morte na Bacia de Campos." March 20. Available at: www.aepet.org.br. Accessed May 2001.

———. 2001b. "Tragédia com a P-36." *Petróleo e Política* (Oil and Politics), 27. Rio de Janeiro. Available at: www.aepet.org.br. Accessed April 20, 2001.

———. 2004. "Discurso de Transmissão da Presidência da AEPET." Fernando Siqueira, February 6, 2004. Available at: www.aepet.org.br/textosprincipal.asp?sqtipotexto=5&dttexto.

Agencia Nacional do Petróleo/Diretoria de Portos e Costas. 2001. "Análise do Acidente com a Plataforma P-36, Julho."

Ahumada, Consuelo. 1996. *El Modelo Neoliberal y su Impacto en la Sociedad Colombiana.* Bogota: Editorial El Âncora.

Amaral, Marina. 2001. "Mais uma da Petrobrax." *Caros Amigos* (São Paulo, Brazil) 5, 49:14–17.

Background Section of the Preamble to the Occupational Safety and Health Administration Process Safety Management of Highly Hazardous Chemicals. 1996. 1. See 29 CFR Part1910 Section 1200. U.S. Government Printing Office (GPO); 1999.

BBC News. 2001. "Brazil Oil Spill Company in Spotlight," March 16. Available at: http://news.bbc.co.uk/hi/english/world/americas/newsid_1225000/1225749.stm. Accessed April 22, 2001.

Brecher, Jeremy, Tim Costello, and Brendan Smith. 2000. *Globalization From Below: The Power of Solidarity.* Cambridge, MA: South End Press.

Chomsky, Noam. 1999. *Profits over People: Neoliberalism and Global Order.* New York: Seven Stories Press.

Coburn, David. 2000. "Income Inequality, Social Cohesion and the Health Status of Populations: The Role of Neo-liberalism." *Social Science & Medicine* 51, no. 1: 135–46.

Druck, Maria da Graca. 1999. *Terceirização: (des) Fordizando a Fábrica: Um Estudo do Complexo Petroquímico.* Salvador, Bahia: EDUFBA and Boitempo Editorial.

Federação Única dos Petroleiros. 2001. "Trabalhadores Mortos em Acidentes de Trabalho na Petrobrás–1998 a 2001." October 31. Available at: www.fup.org.br. Accessed May 2001.

Final Report. 2001. "Inquiry Commission P-36 Accident, Rio de Janeiro, Brazil, June 22." Available at: www.taproot.com/p36_Final_Report.pdf.

Fiori, Jose L. 1997. *Os Moedeiros Falsos.* Petrópolis: Editora Vozes.

———. 1998. "Neoliberalismo e Políticas Públicas." In *Os Moedeiros Falsos.* Rio de Janeiro: Vozes.

Franco, Tania, ed. 1997. *Trabalho, Riscos Industriais e Meio Ambiente: Rumo ao Desenvolvimento Sustentável?* Salvador, Brazil: EDUFBA.

Freitas, Carlos M., Carlos Augusto V. Souza, Jorge M. Machado, and Marcel F. Porto. 2001. "Work-related Accidents on Offshore Oil Drilling Platforms in the Campos Basin." *Cadernos de Saúde Pública* 17: 117–30.

Frumkin, H., and Volney Câmara. 1991. "Occupational Health and Safety in Brazil." *American Journal of Public Health* 81, no. 12: 1620–24.

FUP (United Federation of Oil Workers). 2001. "Sindicato dos Petroleiros Denuncia Precarização do Trabalho." March 9. Available at: www.fup.org.br. Accessed May 2001.

Hora do Povo (São Paulo). AEPET. 2001. "Acidente com a P-36 Não Pode Ser Encarado como uma Fatalidade." Ano XII, August 7. Available at: www.horadopovo.com.br. Accessed September 2001.

IHU On Line (Brazil). 2003. Anon 3, número 86. "Economia Brasileira: Entre os Neoliberais e os Nacionais-desenvolvimentistas." (Brazilian Economy: Between the Neoliberals and the National-developmentalists), December 1. Available at: www.ihu.unisinos.br/boletim/edicoes. Accessed February 2004.

John Gray Institute. 1991. "Managing Workplace Safety and Health: The Case of Contract Labor in the US Petrochemical Industry." Lamar University System.

Jornal do Brasil. 2003a. October 4, A17. Rio de Janeiro.

———. 2003b. October 12, A26. Rio de Janeiro.

Karliner, Joshua. 1997. *The Corporate Planet: Ecology and Politics in the Age of Globalization.* San Francisco: Sierra Club.

O Globo. 2003. December 14, 37. Rio de Janeiro.

Offshore Technology Conference (OTC). 2004. Available at: www.otcnet.org/otcnet/archives/awards_corp.html. Accessed February 21, 2004.

Oficina de Informações. 2001a. "A Reengenharia de Reichstul" (The Reengineering of Reichstul), semana de 13 a 17 de agosto de 2001. Available at: www.oficinainforma.com.br. Accessed February 16, 2004.

———. 2001b. "A Tragédia da Petrobrás" (The Tragedy of Petrobrás), semana de 13 a 17 de agosto de 2001. Available at: www.oficinainforma.com.br. Accessed February 16, 2004.

———. 2001c. "Um Relatório e Suas Entrelinhas" (A Report and Its "Between the Lines" Implications), semana de 13 a 17 de agosto de 2001. Available at: www.oficinainforma.com.br. Accessed February 16, 2004

PACE (Paper, Allied-Industrial, Chemical and Energy Workers International Union). 1999. *Incident Investigation Training.* Nashville, TN: PACE.

Petras, James, and Henry Veltmeyer. 1999. "Latin America at the End of the Millenium." *Monthly Review* 11, no. 3: 31–52.

Petrobrás. 2004. Available at: www2.petrobras.com.br/ri/ingles/apresentacoes/pdf/final_ report0622.pdf. Accessed March 10, 2004.

———. 2004a. Available at: www.petrobras.com.brr/ri/esp/apresentacoeseventos/apresentacoes/pdf/Closing_B ell_NYSE_50anos_ingles.pdf. Accessed March 14, 2004.

———. 2004b. Available at: www2.petrobras.com.br/portal/ingles/meio_ambiente.htm. Accessed March 14, 2004.

Rocha, Geisa. 2002. "Neo-Dependency in Brazil." *New Left Review* 16: 1–33.

Sader, Emir. 2003. "Lula Ano I" (Lula Year I). Available at: www.consciencia.net/2003/12/12/sader1.html. Accessed March 2004.

Sindipetro, Surgente. 2001. Newsletter of the Rio de Janeiro Oil Workers Union, March 4, No. 783. Available at: www.sindipetro.org.br.

———. 2004. "Boletim 856," p. 2. Available at: www.sindipetro.org.br. Accessed February 15, 2004.

Siqueira, Carlos Eduardo. 1998. *Dependent Convergence: The Struggle to Control Petrochemical Hazards in Brazil and the U.S.* Sc.D. diss., University of Massachusetts, Lowell.

———. 2003. *Dependent Convergence: The Struggle to Control Petrochemical Hazards in Brazil and the United States.* Amityville, NY: Baywood Publishing.

Tabb, William K. 2002. *The Amoral Elephant: Globalization and the Struggle for Social Justice in the Twenty-first Century.* New York: Monthly Review Press.

Woolfson, Charles, and Matthias Beck. 2000. "The British Offshore Oil Industry After Piper Alpha." *New Solutions* 10, nos. 1–2: 11–65.

World Environment News. 2001. March 21. Available at: www.planetark.org/avantgo/dailynewsstory.cfm?newsid=10175. Accessed May 2001.

11

Health and Safety at Work in Russia and Hungary

Illusion and Reality in the Transition Crisis

Michael Haynes and *Rumy Husan*

Nowhere has the impact of neoliberal policies of deregulation been more traumatic than in the countries of the former Soviet bloc. When the Soviet bloc collapsed in 1989–91, the hope of workers was that this would lead to improvement in their material living standards, greater freedom, and better conditions at work, including in the area of health and safety. But in economic terms this has not happened save for particular sections of the population. "For the majority," writes Joseph Stiglitz of the former Soviet Union, "economic life under capitalism has been even worse than the old Communist leaders had said it would be" (Stiglitz 2002, 133). Hungary, a smaller exsatellite, presents us with a contrasting example of a more "successful" transition where the scale of the crisis was less and the area of political freedom gained greater. Nevertheless, here too the years since 1989 have been difficult ones for the mass of the population. But how did the programs of market reform and deregulation specifically affect the issue of health and safety in the workplace?

In formal terms it would seem that, at least in Hungary, the gain was unambiguous. Hungary led the way in adjusting its health and safety legislation to the new situation with a new Labor Protection Act (Act XCIII of 1993) whose performance was to be overseen by the National Labor Inspectorate. This was then supplemented by the Hungarian Parliament's later agreement to the Hungarian National Program of Occupational Safety and Health, which set out in detail the need for good health and safety at work and some of the problems being experienced in achieving the agreed aim that "all persons working within the territory of the Republic of Hungary shall have the right to safe, healthy working conditions." This positive legislative base in Hungary was partly a product of the national tradition, partly a reflection of that country's earlier adherence to and role in the International Labor Organization (ILO), and partly the commitment to integration with the European Union (EU). In Russia, the formal situation was and remains much less impressive. Successive presidential and parliamentary declarations made claims about health and safety guarantees but there was no early legislation of the Hungarian type. When the issue arose in the late 1990s, discussion divided between those who argued that whatever its deficiencies the promises of the old Soviet legislation were worth keeping and trying to enforce and the arguments of "liberal reformers" that it was better to have new legislation that offered less but delivered more in practice. Given the scale of the transition crisis in Russia and the distrust of both sides, this debate quickly gained a bitter quality. But in both countries, of course, what is important is less the legislation than the way in which it interacts with the real situation on the ground.[1]

By comparing the experience of the two countries, this chapter shows that although the impact of market reform has differed in both countries, in neither has it resolved the problems of health and safety at work that were apparent under the old regimes.

Health and Safety Under the Old Regime

To make sense of this we need to briefly set the situation on the eve of the transition in Russia and Hungary. In 1989 the Union of Soviet Socialist Republics (USSR) was a superpower. Its antecedents lay in the 1930s when the Stalin regime had begun to drive economy and society forward in an attempt to "catch up and overtake" the West. It never succeeded in coming anywhere near overtaking the advanced countries, but for several decades a process of catch-up seemed to be in place (Haynes and Husan 2002).

In the 1980s, however, a degree of fall-back began to be evident as growth slowed and underpinned a search for reform that led eventually to the political collapse of the system in 1989–91. At this point the level of development was much inferior to the leading states in the West—per capita output was perhaps 30 percent that of the U.S. level and just over 40 percent of that in the United Kingdom.[2] But in global terms this was compensated to some extent by its large population of 287 million and this enabled the USSR to continue to mount a formidable military challenge to the West. This was already apparent in 1945 when the Red Army played a decisive role in the defeat of Hitler. Moreover it was then in a position to incorporate a bloc of Central and Eastern European countries into the Soviet empire as the cold war developed. Hungary, a smaller but more developed country, was one of these. By 1989 it had a population of 10.4 million with a per capita output similar to that in the most advanced parts of the USSR.

Throughout the Soviet bloc the same imperative of competitive industrialization ruled in the years after 1945. The pressure, in part articulated through military competition, was to achieve industrialization, urbanization, and modernization. The means were centralized direction of resources and the economy. However, over time this central direction was modified by the incorporation of more market elements and in 1968 Hungary went furthest in this direction with the implementation of what was called a "New Economic Mechanism."

The official ideology of these countries claimed that they were "Socialist states" that gave pride of place to the interests of the workers who, in turn, were motivated by a "love of the factory" and "pride in labor" to commit themselves to work (Lane 1987, passim). But the experience of work belied these claims and not least in the area of health and safety at work. In Russia in the 1920s there had been an extensive discussion of the health and safety issue and much pioneering work was done. But with Stalin's consolidation of power in 1928–29 the objective now became to catch up to and overtake the West. The interests of the workers took second place to the drive to accumulate to achieve breakneck growth. Opposition to increasing output was seen as counterrevolutionary, and failures, including major accidents at work, were attributed to politically motivated wrecking. In this context there was now no place for statistics of accidents and aggregate health and safety data, which ceased to be published. Even journals devoted to health and safety at work fell victim to the mania for secrecy and the inability to allow the real situation and its causes to be discussed. The health and safety issue was now treated as serious only to the extent that it supported increased production; when it conflicted with increased production it came in a very poor second place. Consequently, laws and regulations were toothless and, without independent means of representation, workers were powerless to enforce them. The situation was, however, rather more liberal in Hungary when the Soviet regime was created there from 1948 onward. Health and safety statistics, for example, continued to be reported, though they were little noticed by outside commentators. But within the workplace similar imperatives existed to those in

Russia. A famous account of work in "socialist Hungary" was Miklos Haraszti's *A Worker in a Worker's State,* for which he was arrested when he tried to get it published in the early 1970s. Here he treated work in Hungary as a form of "wage labor," as in the West, and he showed how piecework pressures operated to encourage workers to "loot"—cut corners, including safety ones, to meet their norms. As in Russia this was formally frowned upon but actively supported in practice: "a man who creates wealth is compelled to work in a way which destroys the quality of that which he produces and his own health as well; simultaneously, he is forbidden to work in this way, as if he would ever choose to do so of his own free will" (Haraszti 1977, 62; see also Burawoy 1985, chap. 4).

The official data now available for 1980 and 1985, the eve of the breakup of the Soviet system, provide a base point for our analysis of the subsequent period.[3] They show a picture that was an improvement on earlier decades but with serious problems in both countries. Because of definitional and other problems it is unwise to make precise cross-country comparisons, but it is safe to say that even if the data in Table 11.1 for these dates are taken at face value, they show a situation in Russia that was much worse than that in Hungary, and this is consistent with what we know from other sources. For the purposes of illustration we can note that when Soviet miners struck in 1989 they drew attention to the fact that the number of miners killed was around a thousand a year and that even something as basic as their soap allowance had not been increased since the 1920s. Beyond this, we also know that across the bloc there was a history of industrial diseases with delayed impacts that continued to be felt, along with high levels of environmental degradation from the heavy-industrial sector that was invested in the attempt to drive up growth at any cost.

The Recorded Pattern of Accidents and Deaths

Table 11.1 shows the official data on accidents and deaths at work in both Hungary and Russia since 1980, but with particular focus on the transition years when the program of liberalization and deregulation had been in place. The numbers of accidents and deaths at work have clearly been reduced by a considerable amount in both countries. But in any comparison over time, or between countries, what matters is not the absolute numbers but the accident and death *rates.* These too show improvement but the absolute numbers and rates do not always move consistently together as we might expect if the statistics were a real reflection of the impact of deregulation on health and safety in the workplace.

Table 11.1

Accidents and Deaths Data for Hungary and Russia, 1980–2001

	Hungary				Russia			
	Accidents	Deaths	Accidents per 1,000 manual workers	Deaths per 100,000 manual workers	Accidents	Deaths	Accidents per 1,000 manual workers	Deaths per 100,000 manual workers
1980	102,017	395	32.8	12.7	570,000	2,349	84	183
1985	116,497	510	39.6	17.3	456,000	9,819	66	142
1989	97,505	440	36.4	16.4				
1990	88,684	428	35.6	17.2	432,000	8,393	66	129
1991	76,325	362	33.3	15.8	406,000	8,032	65	128
1992	55,223	292	29.6	15.7	364,000	7,655	62	131
1993	50,554	243	30.6	14.7	343,000	7,574	63	139
1994	35,919	151	22.4	9.4	300,000	6,770	59	133
1995	33,471	176	22.3	11.7	271,000	6,789	55	138
1996	30,910	151	21.9	10.7	213,000	5,378	61	155
1997	28,896	149	20.9	10.8	185,000	4,734	58	148
1998	28,688	168	20.3	11.9	158,000	4,296	53	142
1999	28,116	161	18.0	10.3	153,000	4,259	52	144
2000	28,220	153	18.0	9.8	152,000	4,404	51	149
2001	26,369	128	16.9	8.2	145,000	4,372	50	150

Source: Hungarian Central Statistical Office, various; Goskomstat Russia, various.

To make sense of the problems in analyzing the pattern of injury and death we can start with the simple idea of the accident/death at work rate, which is given by

$$\frac{\text{Accidents/death at work}}{\text{Number of workers}} \times 1{,}000$$

The numerator represents the total number of officially recorded deaths and accidents. It is well known that the recording of accidents and deaths at work leaves much to be desired. This is not only because of differences over what should be recorded but because, even within agreed definitions, there is a temptation not to record safety violations and their costs properly. Deaths at work and very serious accidents are more likely (but not invariably) to be recorded than less serious accidents. In analyzing how health and safety has changed over time, we have therefore to consider both changes in the real situation and changes in the recording situation since the latter may obscure the former. The problem here is that the very factors that are likely to worsen safety problems are also likely to worsen recording problems. Crudely, we might imagine that managers in a plant affected by difficult economic circumstances might be tempted to encourage workers to cut corners, but they might also reduce the labor force and take a more hostile attitude toward trade unions to help create an atmosphere of fear and insecurity in order to get more out of the workers. In this situation, the temptation will be not to record accidents just at the time when they begin to rise. There is a strong case that this is exactly what happened in both Hungary and Russia in the 1990s. Moreover, if at the same time external checks are disrupted because of changes in government structures and labor inspection, this will only intensify the recording problem.

But problems also exist in regard to calculating the denominator in terms of the nature of the workforce. There are two big problems here. One is the different degrees of vulnerability at work for different parts of the workforce. Over a shorter period of time this will not usually be a problem since workforce structures will tend to change slowly. But if there is an intense period of structural change concentrated into a few years, then any fall in the accident/death rate may simply measure the reduction of workers in more vulnerable sectors rather than a real improvement in workplace safety. The second difficulty is that the denominator refers to workers employed, but if workers are put, say, on short time, or for whatever reason kept on the firm's books after they have ceased to work, this will also reduce the accident/death rate by artificially inflating the size of the labor force. Both of these factors have been at work in Hungary and Russia. Taken together, these suggest a much less optimistic picture than that derived either from the crude numbers of accidents and deaths or the more contradictory picture presented in the data on rates. To see why this is so, we will first consider the impact of the transition and liberalization on conditions in the economies of Hungary and Russia at large, especially as they have been reflected in the structure of the labor force, overall standards of living, and so forth. We will then look more closely at the way in which the structure of vulnerability of workers in the workplace has been affected by the process of liberalization and deregulation in conditions of transition crisis.

The Transition

The political borders of Hungary remained unaffected by the transition, but in 1991 the USSR fragmented into fifteen successor states with the Russian Federation, on which we will focus here, being the largest, with a 1991 population of 148.5 million—just over half the Soviet total.

Figure 11.1 plots the gross domestic product (GDP) level in Russia, Hungary, and the European Union (EU) from 1989 to 2001. It is important to remember that part of the promise of neoliberal reform was that it would allow the catch-up that the old regimes had attempted but failed to deliver. The first thing to note is the catastrophic situation in Russia, where output fell to a low of 56 percent of the 1989 level in 1998. Thereafter a narrow-based recovery took place but still left a peacetime economic collapse without parallel (save in the other fragments of the USSR) in economic history. It was also remarkable compared to some war experiences. Despite the best efforts of the Nazis, for example, Soviet GDP fell by much less in the worst war year. Hungary's experience looks better, but of course the comparison with Russia is extreme. It too experienced a major fall of GDP of 20 percent in the early 1990s. Thereafter, a sustained if uneven recovery took place and in 1999 and 2000 the 1989 level was exceeded for the first time, making it, with Poland, a "success story."[4] But as Figure 11.1 also shows this did not mean a recovery of the 1989 relative position. Hungarian growth roughly tracked that of the EU, so that the gap was still wider than the already wide gap in 1989. Optimistic accounts suggested that the gap might close as a result of the enlargement of the EU. But this is open to doubt. While some of Hungary's regions may benefit, it is questionable if membership alone will prove the basis for a sustained relative improvement in Hungary's position.

Accordingly, for the mass of the population living standards fell, but a small percentage managed to expand their wealth (in some cases spectacularly) and, in consequence, inequality increased in both countries. The costs were borne in both countries by the mass of the population, and poverty

Figure 11.1 **Output Trends in Hungary and Russia, 1989–2001**

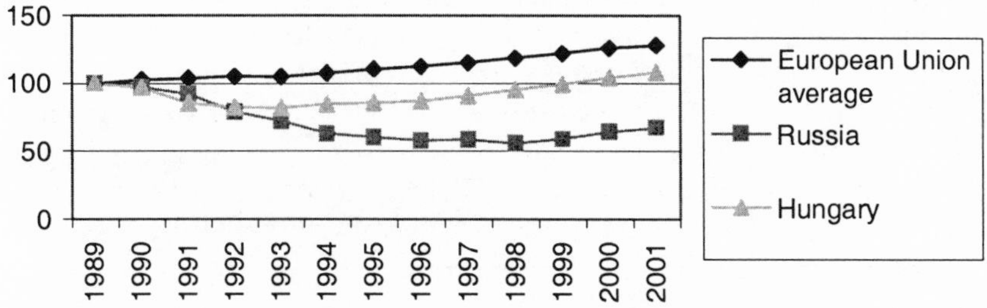

Note: Calculated from data from UN Economic Commission for Europe.

(measured on the rather low official poverty lines) became the lot of a substantial minority. A simple measure of the social strain is that both countries were plunged into a degree of demographic crisis. It was less severe in Hungary, but by 2003 the population, because of the imbalance of birth, deaths, and emigration, was 10 million, less than in 1989. In the Russian Federation the population peaked at 148.6 million in 1992 but had fallen to 144.5 million in 2002 (UNICEF 2002).

However, this hides a much more catastrophic situation because there was major immigration from other parts of the former USSR. Without this, the fall would have been much greater as the death rate jumped and the birth rate sharply contracted. It is difficult to see this trend turning around, even in the medium term, which is why even the more optimistic demographic projections continue to point to population decline.

With this background in mind, let us turn to the issue of the labor force. Figure 11.2 shows the trends in the size of the labor force in the two countries according to the official data. In Hungary 1.5 million workers disappeared from the officially recorded labor force after 1990, but only 400,000 to 500,000 of these were registered unemployed, 800,000 became economically inactive, 150,000 became pensioners, and the number of students rose by 100,000 (Héthy 1999, 3). But when this trend is compared with Russia, a problem is immediately apparent. Despite the Russian economy contracting much more, the labor force seems to have contracted much less than in Hungary. The reason for this is simple. Russia experienced massive disguised unemployment. The Hungarian figures obscure an element of this, but on nothing like the same scale as in Russia, where disguised unemployment became commonplace in the 1990s. Rather than lay people off, many were put on partially paid leave, unpaid leave, and short-time. Some commentators have seen this as a reflection of enterprise paternalism. Gerchikov suggests

Figure 11.2 **Level of Total Employment in Hungary and Russia, 1989–2001**

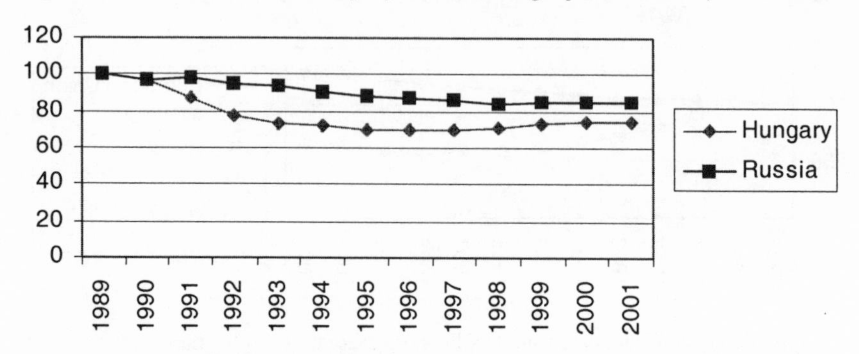

that it equally can be explained as an opportunistic strategy undertaken "simply because it is a cheaper way (in both the moral and material sense) of reducing labour" (Gerchikov 1995, 149).

Whatever the explanation, the existence of this problem means that especially in Russia the labor force data are suspect, and insofar as they increase the size of the labor force this will obviously serve to obscure the true accident level and accident rates.

Economic restructuring also has had a major impact on the sectoral distribution of the labor force and through an increase in the pattern of uneven regional development, which is apparent even in a small country like Hungary. A general trend is evident in both countries and indeed in the transition bloc as a whole. There has been a serious absolute and relative contraction in employment in mining and manufacturing industry and within manufacturing in the heavy industrial sector. In Hungary, for example, manufacturing employment fell by a quarter in a few years and mining by two-thirds. A major contraction also has been evident in agriculture, in which there had been quite high accident and death rates (especially in Russia). There has, however, been a significant expansion in the service sector and in telecommunications.[5] This contraction in the traditional sectors has also affected the relative balance in employment by size of workplace. As is well known, the Soviet bloc had large manufacturing enterprises with huge individual plants, but it was often these that disintegrated during the transition crisis. Consequently, there has been a structural shift toward smaller enterprises and often unregulated forms of microemployment and self-employment including a significant gray economy. The question of uneven regional development is also important in part because the worst problems have always been recorded away from the main centers—"out of sight and out of mind." But uneven development has intensified. Expansion has been concentrated in both the Budapest and Moscow areas. This has created the "Marriott effect"—the

Figure 11.3 **Level of Capital Formation in Hungary and Russia, 1989/1990–2001**

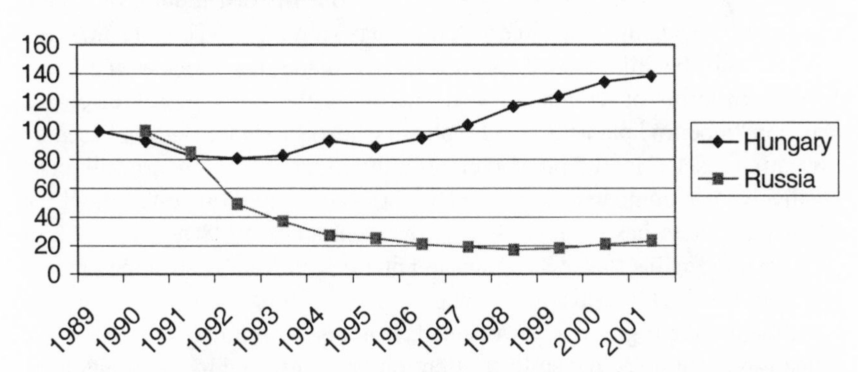

tendency of Western visitors to view the world from the position of their privileged hotels—for insofar as they venture out into the streets the world they see is not that of the mass of the population. It was striking, for example, that on several visits to Moscow for this research it was possible to see that municipal employees engaged in rebuilding and road construction appeared relatively well supplied with safety clothing. But this was part of the mayor's project for improving the city and quite untypical of what could be seen elsewhere in Russia and, indeed, beyond the center of the city. How untypical can be illustrated by the simple figure that by 2000 Moscow, with 7 percent of Russia's population, was generating 20 percent of the Russian GDP. Whereas the national monthly wage was some 4,000 rubles, in Moscow it was nearly 14,000; in the poorest regions it was just over 1,000 (Rossii 2003).

Much the same unevenness is apparent in Hungary. Per capita income in Budapest is nearly twice the national average and three and a half times the poorest county. The Budapest region attracts the lion's share of foreign investment and over half of the foreign-owned firms in Hungary are based there (Hungarian Central Statistical Office 2002, 190–95).

There is one further important general issue for an understanding of the real pattern of health and safety in the new era. Thus is the pattern of capital formation—the aggregate level of which is set out in Figure 11.3. Because the old regimes were engaged in a process of catch-up with the West, the average technical level of their enterprises always lagged behind the average levels of Western Europe, the United States, and Japan. In the last years of the old regimes this lag had begun to increase and this factor encouraged the old leaderships to begin a process of reform. The transition was supposed to resolve this problem by bringing improved investment through foreign direct investment and restructuring the domestic investment effort. Figure 11.3

shows that in Hungary gross capital formation fell but has since reversed, assisted by foreign direct investment (FDI). But the distribution of capital formation reflects the unevenness that we have already noted and some sectors have not benefited at all (Radosevic, Varblane, and Mickiewicz 2003, 53–90). In terms of technical levels, therefore, there is now a greater gap between "modern" plants and older plants whose survival has been based on "sweating assets," including workers. But the situation in Hungary still looks positively rosy compared to that in Russia, where capital investment all but collapsed. There has been some improvement since 1998 but the level of investment remains low, which does not augur well for any sustained recovery. The result is that, save for a few privileged sectors, the age of plants and equipment has progressively worsened. It is one of the major mysteries of transition economics that so little attention has been paid to this issue. How can an economy develop a sustained recovery on the basis of minimal investment? But so far as health and safety considerations are concerned, the crisis in capital formation and accumulation has clearly served to make an already bad situation potentially much worse. "No improvement has been seen in working conditions which are becoming worse because the enterprises have no money to improve them," writes a Russian investigator (Gerchikov 1995, 141).

Vulnerability in the Workplace

This takes us directly to the issue of the situation in the workplace and the way that accidents and deaths are recorded. Investigators of health and safety have theorized that different structures of vulnerability exist for workers that depend on issues like market position, size of firm, degree of managerial power, trade union representation, and so on. But these affect not simply the level of accidents but also how they are recorded. The greater the degree of work place vulnerability then, other things being equal, the greater the degree of underrecording. Here we will concentrate on four internal determinants: the power of the employer/manager, the self-confidence of the worker, the role of the trade union, and the role of state labor inspection. Effectively, the transition has increased the power of managers and employers in the context of a worsening market situation and diminished the potential influence of the other three groups.

We should begin by noting that the poor health and safety situation before the transition was reflected in the widespread adoption of hazard pay that seemed to trade off health and safety against higher wages, bonuses, better food supplies, holidays, and so on, rather than workers being offered both good pay and safer working conditions. This "hazard" pay culture was deeply ingrained, to the frustration of organizations like the ILO, which has been concerned to change the culture of both employers and workers. In Russia, for example, a 1998 survey found that some workers were receiving hazard pay in 94 percent of workplaces and all workers were receiving it in 74 percent (ILO 2000, 39).

This perhaps gives an exaggerated impression of its dimensions, but it clearly shows the problem and the culture of acceptance of a trade-off. In Hungary the role of hazard pay has significantly declined since 1989, but it still exists in some jobs. Enterprises under the old regime were ostensibly run by the state on behalf of "the working class." However, as we suggested at the outset, the real situation was rather different and many commentators now see the transition as, in social terms, involving more a consolidation in a private form of a managerial power over ordinary workers that had previously been articulated through the state. But this does not mean that the new property forms are irrelevant. If the process of privatization effectively consolidated managerial power in the workplace, it also helped to make that power even more unrestrained. The process of privatization was much more ordered in Hungary than in Russia; in the former it was also more influenced by foreign takeover through direct investment. Where this has occurred, it has usually been argued that foreign ownership has brought with it a higher level of technology, better management practices, and higher workplace standards and discipline. By the late 1990s foreign-owned plants in Hungary were producing a quarter of GDP and around one-third of exports (Héthy 1999, 2).

However, while arguments about the relative merits of foreign ownership remain plausible, its actual impact on health and safety remains to be investigated. In the absence of FDI, the restructuring of enterprise control and the creation of new small and medium-sized firms in the context of economic crisis effectively gave local employers and managers much more power. The terms of this created much hostility—not the least in Russia. Gerchikov quotes a worker who said, "Privatization is robbery and an undertaking which makes the people into slaves. Only the top gets something from it. But the top did not operate in the past, as they operate today. As for us, we feel we are slaves. If we object to the present situation we will be fired" (1995, 141). The problem, however, goes beyond the issue of control at the top. In Russia especially, Gerchikov (1995) has suggested that the transition has had a negative impact on line management by devaluing the role of the technical specialist in favor of a climate that has valued the manager as "wheeler-dealer."

If the power of the employer/manager has increased, then, concomitantly, the informal power of the worker has decreased. With its focus on growth at all costs, the old system created a labor shortage economy. This was true even in Hungary, where market reforms operated during the 1980s. Some unemployment existed and workers could be dismissed. However, full employment conditions gave workers some de facto individual bargaining power if only in terms of forcing managers to consider the difficulty of their replacement if they left to take a job at another plant. With the transition and the emergence of mass unemployment and widespread immiseration, this situation changed fundamentally for most workers, and this is reflected in the comment quoted earlier that "if we object to the present situation we will be fired." Surveys of workers have found that job security is a permanent concern and in some parts of the economy overwhelming. In Russian manufacturing in 1998, for example, 51.9 percent of workers put this as their first concern; in the primary sector the figure was 40.6 percent, and in services 35.1 percent (ILO 2000, 14–15).

This fear has helped to accentuate workers' dependence on their jobs and management within the plant. Such dependency has been increased by problems of wage payment, supplies, and so on, and especially so in Russia. The result is that in a sense workers have become even more divided than they were under the old regime and so tied to the plants in which they work. Workers appear from survey data to have little idea of how their workplace compares with a "safe one," suggesting a degree of formal resignation to the conditions they find (ILO 2000, 33). In both Hungary and Russia this has encouraged atomistic individual survival strategies or individual "exit" strategies over those based on a "collective voice." And if this is apparent within plants, it is no less apparent between plants as unevenness and differentiation have sharply increased. The extent of this varies considerably, but in some areas it is extreme, as evidenced from the following remark from a trade union leader in Siberia: "How [can we] call out on strike the workers from transport, power stations, meat or chocolate factories if their wages are now 3–4 times more than in the machine building industry? For so many years we have heard about workers' solidarity, but now it is practically gone!" (Gerchikov 1995, 156).

It is, of course, the weakness of the worker as an individual that lies behind arguments for collective strength in the form of trade union organization. The capacity of workers to have access to independent trade unions that see the protection of health and safety as part of their general role and offer support to individual workers either in contesting problems with management or taking up matters through the courts, has been important in improving conditions throughout the world. All the transition countries now have formally guaranteed workers' rights and the right to form trade unions and strike. Hungary was one of the countries that led the way, with an independent trade union being formed in 1988, the right to strike being given in 1989, and a new labor code in 1992. In the former Soviet Union, the huge mining strikes in 1989 were an important precursor to the formal recognition of the role of independent organizations in the 1990s.

But in both Hungary and Russia trade unions have been in a peculiar position since 1989 and 1991, respectively. There is an appearance of strength, but a reality of great weakness, and not least where it matters—in the workplace. Sometimes this is formulated as a problem of "union decline" from the strong position of "trade unions," under the old regimes. But since these organizations hardly deserved the name "trade union" the situation is more that real unions have been "born weak" in the transition. As the old regimes crumbled in both Hungary and Russia, new independent unions were formed and old "unions" were forced to try to become real unions. Membership fell from the illusory high figures of the past, but it is still respectable in comparative terms and especially so in Russia. Trade unions, as Crowley and Ost point out, are still the largest civic organizations in these countries. They are also relatively asset rich inasmuch as the property of the state unions was transferred, sometimes acrimoniously, to them (Crowley and Ost 2001, 6–7).

Governments have also looked to create nationwide (and in Russia, regional) tripartite structures, which have seemingly given trade unions a role in managing the transition, but the reality is very different.[6] Nationally, tripartism is "more illusory than real" and this is in no small part because at the base unions are so weak. They have found it easier to maintain themselves in older state plants. Their penetration of the private, and more especially small and medium-sized, firms is weak. In Hungary, Tóth has described "the drastic deunionisation [that] took place in the private sector" (Tóth 2001).

The unions have not successfully broken away from the old regime system where they appeared part of the management structure, and in Russia part of their appeal remains that of being a "consumer cooperative society" (Crowley 2001). Workers do not therefore see unions as a selective means of workplace protection nor do they see them as a means of individual defense on issues like health and safety.

The final element we need to consider is the role of the state and the state labor inspection. The old system appeared to provide strong labor protection, with trade unions having a central role. As we have seen, the real situation was less impressive, and already before 1989 Hungary had created an essentially state-based system of health and safety inspection. Since then laws in both countries have changed as labor legislation has been "modernized." These changes have involved the incorporation or influence of ILO conventions and, in the case of Hungary, European Union accession requirements. A new Hungarian labor code was introduced in 1993. In Russia the development of a new labor code involved more protracted and sharper debates as it was seen to risk making an already bad situation worse by removing many of the theoretical obligations that remained from the old era. But the importance of law is not simply what it says but how it is enforced. Here the situation is much better in Hungary than in Russia, but even there we can see an element of formalism that arises both from within and without. To meet EU accession requirements, for example, the Hungarian Parliament had to pass regulations protecting the safety of workers on its deep sea fishing fleet even though it is a landlocked country with no such fleet. Nor is it the case that external requirements are always superior. In the case of Hungary, for example, published accident statistics before 1989 had a remarkable level of detail. This ceased to be the case in the 1990s when the Central Statistical Office (in part because it was not required to report such detail externally) ceased to ask for the data and so the labor inspection organizations ceased to tabulate it.

The crucial question, however, in both countries is how labor inspection is carried out on the ground. The authorities claim that inspectors have appropriate powers that they are not afraid to use when they need them. But we know that in the best of circumstances health and safety inspectors have been accused of preferring a "light touch" even when the actions (or inactions) of employers might justify criminal charges. This is partly because the inspectors are uncertain as to how much support they will get from the courts, but also because there might be a culture of "compromise" where the needs of the employer and the company are rated more highly than the worker. Moreover, this culture of compromise has been argued to increase in a crisis situation. This seems to exist in Hungary and Russia, where inspectors are overworked and can at best inflict relatively small penalties. In Russia too we would have to note a widespread sense that inspection is also affected by a culture of bribery and intimidation. Moreover, just as under the old regime there was a sense that everything had to be shown to be improving, so the same culture still exists under the new regime. This too affects the inspection system and the way it records problems. Any increase in the recording of accidents at work would be seen as a problem even if it were simply a result of a more accurate recording of the real situation.

All of this points to a serious underreporting problem of accidents compared with that in more advanced countries. The scale of this problem is likely to be significantly greater in Russia. There, workers with already low expectations about what should be done, suggested by their survey responses that only 69.8 percent of workplaces reported work-related accidents to the authorities and only 53.2 percent reported work-related illnesses.

The latter figure is especially worrying. The value of the accident as a measure of health and safety is that it is a clear event with a relatively clear cause. Broader health and safety issues are harder to measure and explain in terms of a chain of causes, but in the long run more mortality derives from work-related health problems than accidents. Hungary appears to have a better record here than Russia, and industrial disease statistics, whatever their inadequacies, have been in the public domain since the mid-1980s. But Hungarian standards are still below those of Western Europe. In Russia, conditions do not seem to have improved much at all in working plants. In the ILO survey that we have drawn on, dust and fumes were noted by workers as the single most important problems in manufacturing, but in primary production the survey almost went off the scale; "a staggering 90.9 percent of respondents in this sector considered that dust and fumes were serious problems in their workplaces" (ILO 2000, 14, 16–17).

Conclusion

In modern states workers are offered some protection in law, however inadequate, from risks to their life and health. No such protection exists for whole economies and populations. This is perhaps fortunate because a cynic might suggest that were policies of market deregulation such as those implemented in Russia and Hungary themselves subjected to a national health and safety audit, they would fall at the first hurdle. In the early 1990s such a pessimistic conclusion would have seemed out of place. The collapse of the Soviet-type system in Eastern Europe narrowed down the debate about alternatives. It seemed self-evident that there was only one way forward and although this did not predetermine the exact path, it left limited room for variation. Internal and external critics were marginalized. This was partly because of the stridency with which liberalization was supported and partly by their direct co-option as advisers or through the attraction of a grant culture that rewarded supportive and constructive criticism but looked askance at destructive criticism.

Today such considerations still apply but it is easier to make more far-reaching criticisms even in a supposed "success" story like Hungary. This is not least because the promise of liberalization has not been fulfilled. In too many places, and not least in Russia, it has become mired in its own contradictions that, we have seen, are no less apparent in the area of health and safety at work as elsewhere. The problem is that, as we have seen, registering this is not easy because the very turmoil of liberalization that can produce also serves to obscure much of what is happening on the ground. This is perhaps the final irony. Before 1989 there was a widespread sense among dissidents that not only would the end of the old system bring a material improvement but people would also be able to start "living in truth." Sadly, in the area of health and safety at work, a substantial element of illusion still exists.

Notes

The research for this chapter was funded by an award from The Wellcome Trust: 069459/Z/02/Z. It draws on interviews with the International Labor Organization and its affiliates in Hungary and Russia, the Central Statistical Offices of the two countries, and the Labor Inspection organizations. We are grateful for the time and courtesy shown by all of those involved.

1. Summaries of the key legislative acts for all countries, including Hungary and Russia, are best accessed via the ILO Web site that contains the NATLEX (National Labor, Social Security and Related Human Rights Legislation) database.

2. Comparative data can be found in Maddison (2004).

3. The wider study from which this chapter arises is based on a long-run analysis of health and safety data in Hungary and Russia. We hope to publish our more historical analysis of the earlier period at a later date.

4. However, it is important to point out that personal incomes remained lower and, as Stiglitz suggests, this ought to be the real test of "success" of any program of liberalization (Stiglitz 2002, 121).

5. By 1998, 6.6 percent of Russian GDP was coming from agriculture, 31.3 percent from industry, and 62.1 percent from services (ILO 2000, 4).

6. The nature of value of "tripartism" in the West is a subject of much controversy but in part it has been a concession won by trade unions. In the transition countries it has come from the top down as part of the appropriation of "Western" structures and has even been endorsed by the International Monetary Fund! (See Thirkell, Scase, and Vickerstaff 1995, 177.)

References

Burawoy, Michael. 1985. *The Politics of Production.* London: Verso.

Crowley, Stephen. 2001. "The Social Explosion That Wasn't: Labor Quiescence in Postcommunist Russia." In *Workers after Workers' States: Labor and Politics in Postcommunist Eastern Europe,* ed. Stephen Crowley and David Ost, 199–218. Lanham, MD: Rowman and Littlefield.

Crowley, Stephen, and David Ost. 2001. "The Surprise of Labor Weakness in Postcommunist Society." In *Workers after Workers' States: Labor and Politics in Postcommunist Eastern Europe,* ed. Stephen Crowley and David Ost, 1–15. Lanham, MD: Rowman and Littlefield.

Gerchikov, Vladimir. 1995. "Russia." In *Labour Relations and Political Change in Eastern Europe,* ed. John Thirkell, Richard Scase, and Sarah Vickerstaff. London: University College London Press.

Goskomstat, Rossii. 2003. *Rossiya v Tsifrakh 2003.* Moscow: Gosudarstvennii komitet Rossiiskoi Federatsii po statistike.

Haraszti, Miklos. 1977. *A Worker in a Worker's State: Piece-rates in Hungary.* Harmondsworth, UK: Penguin.

Haynes, Michael, and Rumy Husan. 2002. "Whether by Visible or Invisible Hand: The Intractable Problem of Russian and East European Catch Up." *Competition and Change* 6: 269–87.

Héthy, Lajos. 1999. *Under Pressure: Workers and Trade Unions in Hungary During the Period of Transformation, 1989–1998.* ILO: Budapest Technical Cooperation Project RER/96/M02/NET.

Hungarian Central Statistical Office. 2002. *Statistics of Centuries.* Budapest: Hungarian Central Statistical Office.

International Labor Organization (ILO). 2000. *Trade Union Experiences in Safety and Health at the Workplace in Hungary.* Budapest: ILO-CEET.

———. 2002. *Trade Union Experiences in Safety and Health at the Workplace in the Russian Federation.* Budapest: ILO-CEET.

Lane, David. 1987. *Soviet Labour and the Ethic of Communism: Full Employment and the Labour Process in the USSR.* Brighton, UK: Wheatsheaf Books.

Maddison, Angus. 2004. *The World Economy: Historical Statistics.* Paris: OECD Development Center Studies.

Radosevic, Slavo, Urmas Varblane, and Tomasz Mickiewicz. 2003. "Foreign Direct Investment and Its Effects on Employment in Central Europe." *Transnational Corporations* 12, no. 1: 53–90.

Stiglitz, Joseph. 2002. *Globalization and Its Discontents.* London: Penguin Books.

Thirkell John, Richard Scase, and Sarah Vickerstaff. 1995. "Models of Labour Relations: Trends and Prospects." In *Labour Relations and Political Change in Eastern Europe,* ed. John Thirkell, Richard Scase, and Sarah Vickerstaff. London: University College London Press, 137–68.

Tóth, Andras, 2001. "The Failure of Social-Democratic Unionism in Hungary." In *Workers After Workers' States: Labor and Politics in Postcommunist Eastern Europe,* ed. Stephen Crowley and David Ost, 37–58. Lanham, MD: Rowman and Littlefield.

United Nations Children's Fund (UNICEF). 2002. *Social Monitor 2002.* Florence, Italy: UNICEF Innocenti Research Center.

Contributors

Jordan Barab has worked in the field of workplace safety and health for more than twenty years. He was a former official of the Occupational Safety and Health Administration during the Clinton administration and has spent over seventeen years working in the labor movement. He is currently a recommendations specialist at the U.S. Chemical Safety and Hazard Investigation Board. *Confined Space,* his pioneering blog on occupational safety and health issues, may be viewed at http://spewingforth .blogspot.com.

Peter Dorman is a faculty member in political economy at the Evergreen State College in Olympia, Washington. He is author of *Markets and Mortality: Economics, Dangerous Work, and the Value of Human Life* (1996). Dorman has written extensively on occupational safety and health, child labor, environmental issues, and international trade. He has also consulted for the U.S. labor movement, the U.S. Department of Labor, and the International Labor Organization.

Joan Greenbaum is a professor of computer information systems at LaGuardia Community College (CUNY) and is on the faculty in environmental psychology at CUNY's Graduate Center. She has written extensively about technology and work, and her most recent book is *Windows on the Workplace: Computers, Jobs and the Organization of Work* (2004). Greenbaum and David Kotelchuck are cochairs of Health and Safety for the Professional Staff Congress of CUNY (AFT Local 2334).

Nadia Haiama-Neurohr graduated with a degree in biological sciences from the University of Sao Paulo, Brazil, in 1994. She has a master's and a doctoral degree in Cleaner Production and Pollution Prevention from the University of Massachusetts, Lowell. Haiama-Neurohr is currently a senior policy officer at Greenpeace International, where she is responsible for networking with nongovernmental organizations to promote activities to influence EU decision making; preparing Greenpeace's positions; contacting relevant entities within the EU Institutions industry players; and dealing with media, among other tasks.

Michael Haynes lectures in European studies at the University of Wolverhampton in the United Kingdom. He has written extensively on the comparative development of former Soviet bloc countries and is the author of *Russia: Class and Power, 1917–2000* (2002) and coauthor with Rumy Husan of *A Century of State Murder? Death and Policy in Twentieth-Century Russia* (2003), and with whom he has completed a comparative research project on industrial injuries in Hungary and Russia.

Rumy Husan is a senior lecturer at the Science Policy Research Unit, University of Sussex, UK. Previously he was a senior lecturer in the Development of Emerging Markets with special reference to Eastern Europe and the former Soviet Union at the Leeds University Business School, UK. Husan received his PhD from the University of Oxford in the field of the economics of transition in 1994. He has coauthored two books, one with Michael Haynes titled *A Century of State Murder? Death and State Policy in Twentieth-Century Russia* (2003), and has published several articles in academic journals. They have also cowritten articles for the *Journal of Economic Issues,* the *Journal of European Economic History,* the *Journal of European Area Studies,* and *Competition and Change,* among others.

Penney Kome is an award-winning journalist and author based in Calgary, Canada. Her sixth book, *Wounded Workers: The Politics of Musculoskeletal Injuries,* stirred up controversy in Canada when it appeared in 1998. Dave Bennett, director of Occupational Health and Safety, Canadian Labour Congress, wrote that "this will be the best guide to musculo-skeletal injury for years to come." A national columnist for twelve years and a city columnist for four years, she now works as editor of *Straight Goods,* a weekly online news roundup, at www.straightgoods.com.

David Kotelchuck is an associate professor in the Urban Public Health Program at Hunter College, and is director of Hunter College Center for Occupational and Environmental Health. Since 2000 he has been cochair of the

Health and Safety Committee of the Professional Staff Congress of CUNY. Kotelchuck received his PhD in physics from Cornell University and his MPH in Occupational Health and Safety from Harvard School of Public Health. He is a Certified Industrial Hygienist. For many years he has been a member of the Board of Directors of the New York Committee for Occupational Safety and Health (NYCOSH). During 2003–4, he chaired the Occupational Health Section of the American Public Health Association.

Gerald Markowitz is Distinguished Professor of History at John Jay College of Criminal Justice and the CUNY Graduate Center. He is also an adjunct professor in the Department of Sociomedical Sciences at the Mailman School of Public Health. He and David Rosner have published *Deceit and Denial: The Deadly Politics of Industrial Pollution* (2002). He has been awarded numerous grants, including from the National Endowment for the Humanities, and was a recipient of the Viseltear Prize from the Medical Care Section of the American Public Health Association for "outstanding contributions" to the history of public health. He has coauthored and edited with David Rosner numerous books and articles, including *Deadly Dust: Silicosis and the Politics of Occupational Disease in Twentieth-century America* (1991; 1994); *Children, Race, and Power: Kenneth and Mamie Clark's Northside Center* (1996); *Dying for Work* (1987); and *Slaves of the Depression: Workers' Letters About Life on the Job* (1987).

Vernon Mogensen is an associate professor of political science at Kingsborough Community College of the City University of New York. He received his PhD in political science from the CUNY Graduate School. He is the author of *Office Politics: Computers, Labor, and the Fight for Safety and Health* (1996). Mogensen has written extensively on labor issues and guest edited, and contributed to, a special issue of *WorkingUSA* (Fall 2003) on occupational safety and health issues. His research interests focus on the conflicts between democratic forces and corporate interests in the making of public policy regarding labor and environmental issues.

Rory O'Neill is the editor of *Hazards,* the award-winning UK-based trade union health and safety journal, and is health, safety, and environment officer for the International Federation of Journalists. He also edits a number of online trade union publications.

Laura H. Rhodes, CSP, is an associate professor at Indiana University of Pennsylvania, Safety Sciences Department, where she teaches both undergraduate and graduate courses, recently developing and delivering a dual-level course, "Current Issues in Safety." She is an OSHA-authorized instructor in OSHA standards, where the majority of her teaching load rests. Rhodes also oversees senior interns at major corporations in the United States. She is the adviser to the Safety Sciences Honor Society Rho Sigma Kappa. Her doctoral dissertation from the University of Pittsburgh in Administration and Policy Studies focused on the intersection of the human resources and safety professions. She has published articles in *Professional Safety,* the journal of the American Society of Safety Engineers related to human resources and safety issues.

David Rosner is a professor of history and public health at Columbia University and director of the Center for the History of Public Health at Columbia's Mailman School of Public Health. He and Gerald Markowitz have authored *Deceit and Denial: The Deadly Politics of Industrial Pollution* (2002). He received his BA from City College of New York, MPH from the University of Massachusetts, and his PhD from Harvard in the history of science. Before moving to Columbia in 1998 he was University Distinguished Professor of History at the City University of New York. In addition to receiving numerous grants, he has been a Guggenheim Fellow, a National Endowment for the Humanities Fellow, and a Josiah Macy Fellow. Presently he is the recipient of a Robert Wood Johnson Investigator Award. He has been awarded CUNY's Distinguished Scholar's Prize and, recently, the Viseltear Prize for Outstanding Work in the History of Public Health from APHA and the Distinguished Alumnus Award from the University of Massachusetts.

Carlos Eduardo Siqueira graduated with a degree in medicine from the Medical School of the Federal University of Rio de Janeiro in 1979. He has a master's in public health from the Johns Hopkins University School of Hygiene and Public Health and a doctoral degree in Work Environment Policy from the University of Massachusetts, Lowell. Siqueira is a research assistant professor in the Department of Work Environment at the University of Massachusetts, Lowell, where he has researched the political economy of the migration of hazards between developed and developing countries, and health care and immigrant workers' work environment policy issues. He published the book *Dependent Convergence: The Struggle to Control Petrochemical Hazards in Brazil and the United States* (2003).

Robert Storey teaches in the Labour Studies and Sociology departments at McMaster University in Hamilton, Ontario, Canada. He is a founding member of Canada's premier worker history museum, the Workers' Arts and Heritage Centre, and is currently a board member of the Occupational Health

Clinics for Ontario Workers. Before changing his primary research focus to worker health and safety, Storey published historical and contemporary articles on workers and unions in the Canadian steel industry. (He also worked in Hamilton's steel mills before and while going to university.) Presently, he is concentrating his studies on the genesis and evolution of health and safety and, most recently, injured-worker activism in Ontario. Like the individuals he has interviewed in his research, he believes—passionately—that workers' health is not for sale.

Eric Tucker is a professor at Osgoode Hall Law School, York University, where he teaches labor law and occupational health and safety regulation. He is the author of *Administering Danger in the Workplace: The Law and Politics of Occupational Health and Safety Regulation in Ontario, 1850–1914* (1990) and coauthor (with Judy Fudge) of *Labour Before the Law: The Regulation of Workers' Collective Action in Canada, 1900–1948* (2001) and (with Cynthia Cranford, Judy Fudge, and Leah Vosko) of *Self-Employed Workers Organize* (2005). Tucker has written extensively on occupational health and safety regulation, focusing on worker rights and enforcement issues, and has worked with health and safety activists and injured-workers' groups in Ontario.

Index